London & North Eastern Railway 4-4-0 Tender Locomotives

Front cover photo:
62264, one of the first batch of Pickersgill 'V' class built by Neilson & Co. in 1899, later LNER D40, at Elgin shed, 21 May 1956. (Transport OnLine Collection)

Back cover photos:
D20/1 62387 at Alne with the RCTS 'Yorkishire Coast Rail Tour', 23 June 1957. (Transport OnLine Collection)

D49/2 235 *The Bedale* at York, 1938. (Colour Rail)

London & North Eastern Railway 4-4-0 Tender Locomotives

North Eastern, North British, Great North of Scotland, L N E R

DAVID MAIDMENT

First published in Great Britain in 2025 by
Pen and Sword Transport
An imprint of Pen & Sword Books Ltd.
Yorkshire - Philadelphia

Copyright © David Maidment, 2025

ISBN 978 1 39903 684 9

The right of David Maidment to be identified as author of this work has been asserted by him in accordance with the Copyright, Designs and Patents Act 1988.

A CIP catalogue record for this book is available from the British Library.

All rights reserved. No part of this publication may be reproduced or transmitted in any form or by any means, electronic or mechanical, including photocopy, recording or any information storage and retrieval system, without the prior written permission of the publisher, nor by way of trade or otherwise shall it be lent, re-sold, hired out or otherwise circulated without the publisher's prior consent in any form of binding or cover other than that in which it is published and without a similar condition including this condition being imposed on the subsequent purchaser.

Typeset in Palatino by SJmagic DESIGN SERVICES, India.
Printed and bound by Printworks Global Ltd, London/Hong Kong.

Pen & Sword Books Ltd. incorporates the imprints of Pen & Sword Books: After the Battle, Archaeology, Atlas, Aviation, Battleground, Discovery, Family History, History, Maritime, Military, Politics, Select, Transport, True Crime, Fiction, Frontline Books, Leo Cooper, Praetorian Press, Seaforth Publishing, Wharncliffe and White Owl.

For a complete list of Pen & Sword titles please contact

PEN & SWORD BOOKS LIMITED
George House, Units 12 & 13, Beevor Street, Off Pontefract Road,
Barnsley, South Yorkshire, S71 1HN, England
E-mail: enquiries@pen-and-sword.co.uk
Website: www.pen-and-sword.co.uk

or

PEN AND SWORD BOOKS
1950 Lawrence Rd, Havertown, PA 19083, USA
E-mail: uspen-and-sword@casematepublishers.com
website: www.penandswordbooks.com

All David Maidment's royalties from this book will be donated to the Railway Children charity [reg. no. 1058991] [www.railwaychildren.org.uk]

Other books by David Maidment:
Novels (Religious historical fiction)
The Child Madonna, Melrose Books, 2009
The Missing Madonna, PublishNation, 2012
The Madonna and her Sons, PublishNation, 2015
The Reluctant Traitor, PublishNation, 2021

Novels (Railway fiction)
Lives on the Line, Max Books, 2013
Steamy Stories, PublishNation, 2021

Non-fiction (Railways)
The Toss of a Coin, PublishNation, 2014
A Privileged Journey, Pen & Sword, 2015
An Indian Summer of Steam, Pen & Sword, 2015
Great Western Eight-Coupled Heavy Freight Locomotives, Pen & Sword, 2015
Great Western Moguls and Prairies, Pen & Sword, 2016
Southern Urie and Maunsell 2-cylinder 4-6-0s, Pen & Sword, 2016
Great Western Small-Wheeled Double-Framed 4-4-0s, Pen & Sword, 2017
The Development of the German Pacific Locomotive, Pen & Sword, 2017
Great Western Large-Wheeled Double-Framed 4-4-0s, Pen & Sword, 2017
Great Western Counties, 4-4-0s, 4-4-2Ts & 4-6-0s, Pen & Sword, 2018
Southern Maunsell Moguls and Tank Engines, Pen & Sword, 2018
Southern Maunsell 4-4-0s, Pen & Sword, 2019
Great Western Granges, Pen & Sword, 2019
Cambrian Railways Gallery, Pen & Sword, 2019
Great Western Panniers, Pen & Sword, 2019
Great Western Kings, Pen & Sword, 2020
Great Western & Absorbed Railway 0-6-2Ts, Pen & Sword, 2020
Drummond's L&SWR Passenger & Mixed Traffic Locomotives, Pen & Sword, 2020
Southern 0-6-0 Tender Locomotives, Pen & Sword, 2021
LNER 4-6-0 Locomotives, Pen & Sword, 2021
Midland & LMS 4-4-0s, Pen & Sword, 2021
Great Western Castle 4-6-0 Locomotives, 1923-1959, Pen & Sword, 2022
Great Western Castle 4-6-0 Locomotives, The Final Years 1960-1965, Pen & Sword, 2022
Great Western Castle 4-6-0 Locomotives, In the Preservation Era, Pen & Sword, 2023
Four-coupled Tank Locomotives, Built by the Great Western Railway, Pen & Sword, 2023
Four-coupled Tank Locomotives, Absorbed by the Great Western Railway, Pen & Sword, 2023
The Princess Coronation Pacific Locomotives, 1937-1956, Pen & Sword, 2023
The Princess Coronation Pacific Locomotives, The Final Years & Preservation, Pen & Sword, 2023
Great Western 0-6-0 tender locomotives, Pen & Sword, 2024
LNER 4-4-0s of Great Northern, Great Central & Great Eastern Design, Pen & Sword, 2024

Non-fiction (Street Children)
The Other Railway Children, PublishNation, 2012
Nobody ever listened to me, PublishNation, 2012

CONTENTS

Preface & Acknowledgements 7

Introduction 8

Chapter 1   **The Engineers** 9
           Alexander McDonnell 9
           Thomas William Worsdell 9
           Wilson Worsdell 9
           Matthew Stirling 10
           Dugald Drummond 10
           Matthew Holmes 11
           William Paton Reid 11
           Walter Chalmers 11
           William Cowan 11
           James Manson 11
           James Johnson 12
           William Pickersgill 12
           T.E.Heywood 12
           Nigel Gresley 12
           Edward Thompson 13

Chapter 2   **The North Eastern 4-4-0s** 14
           The NER class 38 14
           The NER M & Q (LNER D17/1 & D17/2) 17
           The D18 27
           The D19 28
           The D20 30
           The D21 49
           The D22 55
           The D23 59
           The D24 62

Chapter 3   **The North British 4-4-0s** 65
           The D25 65
           The D26 69
           The D27 71
           The D28 71
           The D29 74
           The D30 80
           The D31 93
           The D32 100
           The D33 104
           The D34 106
           The D35 116
           The D36 117

Chapter 4	**The Great North of Scotland 4-4-0s**	119
	The D38	119
	The D39	121
	The D40	122
	The D41	132
	The D42	141
	The D43	144
	The D44	146
	The D45	147
	The D46	147
	The D47	149
	The D48	151
Chapter 5	**The LNER D49 'Shire' & 'Hunt' 4-4-0s**	152
Chapter 6	**Personal Experiences**	181
Chapter 7	**Preserved Locomotives**	186
	NER 'M' 1621	186
	256 *Glen Douglas*	187
	49 *Gordon Highlander*	189
	246 *Morayshire*	191
	Colour Section	193
	Appendix	209
	Bibliography	232
	Index	233

PREFACE & ACKNOWLEDGEMENTS

This is the second of two books that I have written about the LNER 4-4-0s, following up my earlier Pen & Sword book on the LNER 4-6-0s. The majority of the 4-4-0s were designed and constructed by the railways that formed the constituent parts of the LNER in 1923. Only the Gresley D49s (the Shires and the Hunts) were designed and constructed in the LNER era, although the Great Central D11/2 (Directors) were constructed in 1924 for the LNER's Scottish Division. Because of the sheer number of classes to be covered I am writing two volumes for Pen & Sword, the first of which, already published, covered the 4-4-0s of the Great Northern, Great Central and Great Eastern Railways, plus a short chapter on the Midland and Great Northern locomotives that passed to LNER stock in October 1936. This volume covers the 4-4-0s of the North Eastern, North British and Great North of Scotland (GNoS) Railways plus the Gresley D49 designs.

I acknowledge the considerable help of Paul Shackcloth of the Manchester Locomotive Society, who presides over its huge archive of photographs and allows me to include them for publication without any fee as the royalties from the book will be donated, as with my royalties from other books I have written, to the Railway Children charity (www.railwaychildren.org.uk) which I founded in 1995 and supports street and runaway children around the railway and bus stations of India, East Africa and the United Kingdom. The archive contains so many photographs of LNER 4-4-0s that I have hardly needed to search elsewhere. I acknowledge a few from John Scott-Morgan's collection to fill the gaps plus a few early ones I took myself to illustrate my limited personal experience of living and working with these locomotives.

I also acknowledge access to the locomotive performance logs of the Railway Performance Society and their contributors, and their records culled from early articles in the *Railway Magazine* by gurus such as Cecil J. Allen. I have endeavoured to give credit in the captions to the original photographer where known but many of the photos I've have scanned from the MLS collection do not identify the photographic source. If I have missed anyone, please contact the publisher so I can, if possible, make amends.

I thank in particular my editor, Carol Trow and Pen & Sword's Production Manager, Janet Brookes, for their very professional support and the whole of the Pen & Sword team, designers, marketing and production staff for the consistent quality of the books they produce on my behalf.

David Maidment
2025

INTRODUCTION

Passenger traffic in the late nineteenth century grew rapidly and the various railway companies, which had relied on single-wheelers for the majority of their existence were being forced to look at ways of coping with heavier train loads. The famous Stirling and Johnson 4-2-2s and the GW Dean engines were still on main line work, but their loads were limited. The first 4-4-0s of any consequence appeared on the Midland in the late 1870s and a few classes appeared on various railways in the 1880s, but it was the 1890s when the 4-4-0 wheel arrangement flourished on all the main line railway companies of the pre-Grouping age. The first decade of the twentieth century saw the introduction of 4-4-0 classes that reigned on the main express services until the 4-6-0 or 4-6-2s became more widespread. The most famous and most competent were the Great Weatern (GW) 'Cities', the London and North Western Railway (LNWR) 'George Vs', the Great Central (GC) 'Directors', the North Eastern (NE) 'Rs', the Midland 'Compounds', the Caledonian 'Dunalastairs', the Great Eastern (GE) 'Claud Hamiltons', the South Eastern and Chatham Railways (SE&CR) 'Ds' and 'Es' and the London and South Western Railway (LS&WR) 'T9s'. Only on the GW and GE were they replaced on the best expresses by 4-6-0s within their first decade. The Great Central, Northern Eastern and North British produced 4-4-2s for express work in a similar timescale which shared top link work with the 4-4-0s. Only the Great Northern constructed the 4-4-2 'Atlantic' as its prime express locomotive in the first decade of the twentieth century, with its 4-4-0s taking mainly secondary roles.

At the Grouping, the LNER received 136 Great Northern, 146 Great Central, 169 Great Eastern, 186 North Eastern, 183 North British and exactly 100 Great North of Scotland 4-4-0s, 920 in total. Of these, 431 passed into British Railways stock in 1948 plus the 76 built by Gresley – the D49s. Interestingly, the percentage of the pre-Grouping designs extant at nationalisation were Great Northern 42%; Great Central 32%; Great Eastern 62%; North Eastern 28%; North British 52%; Great North of Scotland 40%. The only completely new 4-4-0 designs produced by the 'Big Four' were the LNER D49 'Shire' and 'Hunt' classes built mainly for secondary work in the North East and Scotland and the Southern 'Schools' built initially for the Hastings route and then both Eastern Section mainlines to the Kent Coast via Chatham and Ashford. The 'Schools' were almost certainly the most powerful and competent of all the 4-4-0 British designs.

I have covered the GW 4-4-0s, the Midland 4-4-0s and the Southern 4-4-0s in previous Pen & Sword 'Locomotive Portfolios' and my book on the LNER 4-6-0s was published in 2021. I have now decided that it was time to tackle the LNER's 4-4-0 classes inherited from its constituent companies in a similar fashion and my two volumes cover all classes designated D1 to D54 – all 4-4-0 tender engines apart from a few locomotives of class D50 and D51, 4-4-0 tank engines built by the North British Railway which I excluded. This second volume will cover the design, construction, history, operation and performance of the 4-4-0 designs of the North Eastern Railway, the North British, the Great North of Scotland and the Gresley D49s plus the single Thompson D49 rebuild. I have endeavoured to find many previously unpublished photographs, both detailed portraits of especial interest for modellers and images of the classes in action to illustrate their operation. Information about the operation and particularly the performance of some of the less well-known classes has been hard to come by, but I have included whatever I could find from my researches.

Chapter 1
THE ENGINEERS

Alexander McDonnell, 1883–84, North Eastern Railway

Alexander McDonnell was born in Dublin on 18 December 1829 and was educated there, obtaining a BA in Maths at Trinity College. He was apprenticed at Newell & Gordon, Westminster, and subsequently became an engineer with the Newport, Abergavenny and Hereford Railway. In 1864 he moved back to Ireland to work for the Great Southern & Western Railway charged with bringing order to the Inchicore Works and to attempt to standardise the company's locomotives. He was made Locomotive, Carriage & Wagon Superintendent and remained in that role for nearly twenty years, introducing practices he had seen at Crewe and gained the reputation of recognising and encouraging talent, identifying and encouraging such men as Aspinall, H.A. Ivatt and R.E.L. Maunsell.

At the end of 1882, he moved to Gateshead as the North Eastern Railway's Locomotive Superintendent following Edward Fletcher's retirement. Despite his reputation in Ireland, he was not a success in his new role, train crews being resistant to changes in footplate design, and the first 4-4-0s he built to replace Fletcher 2-4-0s on the main line disappointed. Although they appeared larger than the Fletcher engines, they scarcely equalled their power and were soon relegated to secondary work being unfit to time the main east coast expresses. His goods engines were equally underpowered, and he proffered his resignation to the NER Board in September 1884. He was granted a year's pay indicating perhaps that faults were not his alone and Henry Tennant headed a Locomotive Committee for a few months before the appointment of Thomas Worsdell.

He died in Holyhead on 14 December 1904 aged almost 75.

Thomas William Worsdell, 1885–90, North Eastern Railway

Thomas William Worsdell was born on 14 January 1838 in Liverpool, the eldest son of Nathaniel, part of a large Quaker family. In 1847, he became a boarder at Ackworth, a Quaker school in Yorkshire. He was apprenticed under Ramsbottom at Crewe Works and in 1865 went to the Pennsylvania Railroad in the USA. In 1871, he was invited to return to Crewe by Webb, the LNWR Locomotive Engineer, and after various positions there became Locomotive Superintendent of the Great Eastern Railway in 1881.

His tenure there was short as in 1885 after the McDonnell resignation and the temporary Tennant Locomotive Committee, he was brought to Gateshead as Locomotive Superintendent of the North Eastern Railway. However, his health deteriorated, and he retired for health reasons aged just 52, succeeded by his younger brother, Wilson. Despite his early retirement he survived for another 25 years, dying at Arnsdale on 28 June 1916.

Wilson Worsdell, 1890–1910, North Eastern Railway

Wilson Worsdell was from a Quaker family, the tenth child, fourth son, of Nathanial and Mary Worsdell and younger brother of William Worsdell. He was born on 7 September 1850 at Monks Coppenhall in Crewe and was a pupil at the Friends' School at Ackworth in 1860. He left in 1867 and spent the first six months of that year in the Crewe Works Drawing Office. In July, he went to America and became an engineering pupil working for Edward Williams, Superintendent of Motive Power and Machinery at the Pennsylvania Railroad's Altoona Works. He returned to Crewe in 1871 in the Works under Francis Webb. He got experience in both the erecting shop and drawing office and was appointed assistant foreman at Stafford locomotive shed in 1874. He moved to Bushbury as foreman in 1876 and was in charge of the Chester depot in 1877.

He married Mary Elizabeth Bradford in 1882 and their son Geoffrey was born the following year, who broke the Quaker tradition by being educated at Charterhouse public school. Worsdell left Chester in March 1883 with a glowing testimonial from his depot staff and went as Assistant Mechanical Engineer of the North Eastern Railway at Gateshead. The Chief, Alexander McDonnell, was a controversial character and resigned at short notice in 1884 and his brother William became Locomotive Superintendent. Wilson was re-designated Assistant Locomotive Superintendent of the Northern Division in 1885 and his salary was increased by £100 to £600 a year. William retired on health grounds in 1890 and Wilson took over on 1 October at a salary of £1,100. He remained in charge until May 1910. He bought a home, Greenesfield House, adjacent to the Works, working closely with an outstanding General Manager, George Gibbs. Wilson's assistant was Vincent Raven and Walter Smith was his Chief Draughtsman.

He designed the M1 4-4-0 in 1895, an engine involved in the 1895 East/West Coast 'Race to Aberdeen' and went on a study tour to the USA in 1901. Influenced by what he saw, he designed larger 20 ton wagons and the outstanding Q6 0-8-0s and J27 0-6-0s which remained as last survivors of steam in the North East. He designed the enlarged 4-4-0 of class R (LNER D20) in 1899 and his S 4-6-0s were the first passenger 4-6-0s in the country.

He retired in May 1910 as Chief Mechanical Engineer, his duties including the design and maintenance of locomotives, carriage and wagons, tugboats and hydraulic machinery, having been responsible for the management of 18,500 staff, 2,142 locomotives, 4,000 passenger coaches and 98,000 goods wagons. He was involved in the electrification of North Tyneside in 1904 and the goods branch to Newcastle Docks and electric locomotive 26500, now preserved. He was well liked by his staff, ran a boys' club and became a JP in 1907. He was a keen fisherman, possessing a second house in Voss in Norway, and was President of the Association of Railway Engineers and Carriage & Wagon Superintendents of Great Britain and Ireland. He was much appreciated and remained as a consultant to the North Eastern Railway to the end of 1910. He moved home to South Ascot and died in 1920 aged 69 and was buried at All Souls, Ascot. He was honoured by the naming of a new Peppercorn A1 pacific 60127 *Wilson Worsdell* on 30 October 1950.

Matthew Stirling, 1885–1922, Hull and Barnsley Railway

Matthew Stirling, the son of Great Northern Locomotive Superintendent, Patrick Stirling, was born at Kilmarnock on 27 November 1856. He was apprenticed in his father's Works at Doncaster and worked subsequently for the GNR in its Nottingham District.

He was appointed Locomotive Superintendent, later renamed Chief Mechanical Engineer, of the Hull & Barnsley Railway in May 1885 and held that position until his retirement in 1922 when the North Eastern took over the system for just a year before the Grouping. He died on 5 October 1931 at Kingston-Upon-Hull.

Dugald Drummond, 1875–82, North British Railway

Dugald Drummond was a Scot, born at Ardrossan in Ayrshire on the first day of the year 1840. His father was a railway civil engineer and initially he was apprenticed to the Glasgow engineering firm of Forest & Barr. After spending time in the boiler shop of Thomas Brassey at Birkenhead he moved to Cowlairs Works in 1864 and then foreman erector at Lochgorm Works at Inverness on the Highland Railway. Just three years later aged only 27, he was made Works Manager. Following Stroudley's move to Brighton in 1870 he was in charge of the Highland Railway's locomotive department. After a spell at Brighton with Stroudley, he was appointed Locomotive Superintendent of the North British Railway in 1875 and remained there for seven years where he designed seven new classes of locomotive including the '476' class of 4-4-0s (later LNER D27 and D28) and the class '157' 0-4-2Ts, later modified to 0-4-4Ts (LNER G8).

In 1882 he moved to the St Rollox Works of the Caledonian Railway and between then and 1890 designed and built nine classes for that company, including the class '66' 4-4-0 and the class '171' 0-4-4T, developed by his successors for heavier suburban traffic. He resigned in 1890 and attempted to develop a locomotive construction company in Australia, but it failed and he returned to Britain trying his hand at a similar scheme in Glasgow. He returned to the railway when offered the post of Locomotive

Engineer of the London & South Western Railway in 1895, which he retained until his death in 1912.

In 1909 he designed and built the modern Eastleigh Works to replace Nine Elms and was well known for travelling widely throughout the system, using a unique 4-2-4T nicknamed 'The Bug' which he designed for his personal use. Drummond was a larger than life character, formidable in appearance and temperament, demanding high standards from himself and his workforce. On one of his many forays onto the system he got cold and wet and managed to scald himself in the hot bath he took to restore circulation to his numbed feet. He apparently neglected the burns which turned septic and developed gangrene resulting in the amputation of the leg, dying of shock from the operation, for which he refused an anaesthetic. He was buried at the Brookwood Cemetery served by the Necropolis Railway from Waterloo, whose funeral trains were usually hauled by one of his own 0-4-4Ts.

Matthew Holmes, 1882–1903, North British Railway

Matthew Holmes was born in Paisley in 1844, the son of a businessman. Because of the depression and stagnation of industry there in the mid-nineteenth century, he did not follow his father into the business, but, after education in Edinburgh, he left school at 15 and was apprenticed to Hawthorne & Co. in Leith. He then joined the Edinburgh and Glasgow Railway and after various appointments on the North British Railway, he was by 1873 Foreman at Haymarket. In 1875 he was Chief Inspector for the NBR and Assistant to Dugald Drummond at Cowlairs Works.

In 1882 he succeeded Drummond as Locomotive Superintendent, in charge of 7,000 men. He was described as a tall wiry man of fresh complexion and sharp mind. He was married and lived at Lenzie. He retired in June 1903 and died just a month later, aged 59.

William Paton Reid, 1903–19, North British Railway

William Reid was born in Glasgow on 8 September 1854, the son of Robert Whyte Reid. He was apprenticed at Cowlairs Works in 1879 and by the early 1880s was in charge of the locomotive depot at Balloch. He stepped up to Dunfermline in 1889 and Dundee in 1891 and in 1900 was depot manager at St Margaret's Edinburgh. A new post was then created of Outdoor Running Assistant Locomotive Superintendent which he held until he took on the Locomotive Superintendent's post on the retirement and subsequent early death of Matthew Holmes in 1903. He retired at 65 and died on 2 February 1932, aged 75.

Walter Chalmers, 1920–22, North British Railway

Walter Chalmers was born in 1874, the son of Robert Chalmers who was Assistant Locomotive Engineer at Cowlairs. He followed his father into the employment of the North British Railway and by 1904 was Chief Draughtsman at the Cowlairs Works. He became Chief Mechanical Engineer when Reid retired in 1919, although his official appointment was not until 1920. His tenure of the office however was only two years up to the Grouping when he was appointed to a post in the LNER's Scottish Division. He retired aged 60 in June 1924 and died in 1940. The Chalmers family clearly had an engineering history for his brother, Robert, became Chief Mechanical Engineer of the Queensland Railway in Australia.

William Cowan, 1857–83, Great North of Scotland Railway

Little information is available about William Cowan, the Locomotive Superintendent of the Great North of Scotland Railway between 1857 and 1883. He introduced the first British outside cylinder 4-4-0 and when he resigned from the GNoS post, became a steel tyre salesman for the German firm Krupp as their representative in both the UK and the USA.

James Manson, 1883–90, Great North of Scotland Railway

James Manson was born in 1845 at Saltcoats, Ayrshire and was apprenticed in Kilmarnock Works for eight years. In 1869 he spent a year in the employment of Barclay Curle & Co., shipbuilders of Govan and then was Chief Engineer for ships of the Bibby Line, being at sea between 1870 and 1875. He rejoined the Glasgow and South Western Railway in 1875 as an inspector, then Assistant Works Manager, and by 1878 he was the Works Manager.

He moved to the Great North of Scotland Railway as their Locomotive Superintendent in 1883 where, among other things, he had constructed some 8-wheel tenders, the first in Great Britain. In 1890 he went back to the G&SWR as that company's Locomotive Superintendent and remained there until his retirement in 1911. He was elected a member of the Institute of Mechanical Engineers in 1891, He died at Deanhill, Kilmarnock on 5 June 1935, aged 90.

James Johnson, 1890–94, Great North of Scotland Railway

James Johnson was the son of Samuel Johnson, the Midland Railway Engineer, but little more is known of him other than he was Locomotive Superintendent of the Great North of Scotland Railway between 1890 and 1894.

William Pickersgill, 1894–1914, Great North of Scotland Railway

William Pickersgill was born in Nantwich, Cheshire in 1861. As a Whitworth Exhibitioner he started an apprenticeship at the Great Eastern Works at Stratford in 1876 and worked in many different departments. He rose to the position of District Locomotive Superintendent Norwich in 1891 and succeeded James Johnson as the Locomotive Superintendent of the Great North of Scotland Railway in 1894.

He set up the new workshops at Inverurie replacing the outdated shops at Kittybrewster and was appointed Chairman of the Association of Railway Locomotive Engineers (ARLE) in 1912. In 1914, he succeeded John McIntosh as Locomotive, Carriage & Wagon Superintendent of the Caledonian Railway and after the Grouping, became the Mechanical Engineer of the Northern Division of the LMS, a position he held to 1925 when he retired. He died on 2 May 1928, aged 67.

T.E. Heywood, 1914–22

Thomas Edward Heywood was born at Cardiff on 29 November 1877 and trained under Tom Hurry Riches on the Taff Vale Railway. He won a Whitworth Exhibition gold medal for engineering in 1899 and became a draughtsman, then inspector at Cardiff. In 1902, he emigrated to become the Assistant Locomotive, Carriage & Wagon Engineer of the Burma Railway Company, returning to the UK in 1914 as Assistant Superintendent of Penarth Dock before almost immediately being appointed Locomotive Superintendent of the Great North of Scotland Railway. At the Grouping he was made Running Superintendent of the Northern Scottish Area of the LNER, retiring in 1942. He died in Aberdeen in November 1953.

Nigel Gresley, 1911–41, Great Northern & London North Eastern Railways

Nigel Gresley was born in Edinburgh on 19 June 1876 where his mother had gone for medical treatment due to ante-natal complications, although his father was the Reverend Nigel Gresley, rector of Netherseal. He was their fifth child, the family originally coming from Gresley in Derbyshire. He attended a preparatory school in Sussex and then Marlborough College.

After leaving school he began an apprenticeship under F.W. Webb at Crewe Works, then spent a year undertaking practical work as an 'improver' in the fitting and erecting shops at Crewe. He moved in 1898 to the Design and Drawing Office of the Lancashire and Yorkshire Railway at Horwich under the guidance of John Aspinall. He had a short spell as foreman at Blackpool running shed and ran the materials test room at Horwich. In 1901, he was appointed as the Outdoor Assistant in the Company's Carriage & Wagon Department and the following year, Assistant Works Manager at the Newton Heath depot, promoted to be Works Manager in 1903. Further rapid promotion followed in 1904 when he became Assistant Superintendent of the Lancashire and Yorkshire (L&Y) Railway's Carriage & Wagon Department.

In 1905, he moved to the Great Northern Railway as Assistant Carriage & Wagon Superintendent at Doncaster and in 1906 assumed its leadership. He succeeded H.A. Ivatt as the GN's Chief Mechanical Engineer in 1911 and in 1920 was awarded the CBE for his wartime work at Doncaster. In 1923 J.G. Robinson was offered the senior post of CME of the new LNER Company but declined as he was near retirement and recommended Nigel Gresley who was duly appointed in 1923. He had designed and built the first engines of his 'Big Engine' policy in 1920 (the K3 mogul) and 1922 (the A1 Pacific) but delegated many of the lesser designs to be developed by his Drawing Office staff or contractors under his direction and supervision, including that of the B17.

During the 1930s he lived at Salisbury Hall, near St Albans, and had land where he developed his interest in breeding wild birds and ducks (the inspiration for the names of his A4 Pacifics?). In 1936, he designed the 1,500v D.C. electric locomotives for the Woodhead electrification scheme, although its implementation was delayed until after the Second World War. He was knighted in the 1936 New Year Honours after the triumph of the

'Silver Jubilee' high speed express and the exploits of his Pacifics.

He died suddenly in office on 5 April 1941 and was buried at his father's church of St Peter's, Netherseal, Derbyshire.

Edward Thompson, 1941–46, London North Eastern Railway

Edward Thompson's father was Assistant Master and a Governor of Marlborough School, and a Greek scholar. Edward had three sisters and in 1889 was sent to a preparatory school in Reigate, moving to Marlborough in 1894. Gresley was also a pupil there, two or three years his senior. Edward left in 1899 and took a mechanical science degree at Pembroke College Cambridge, graduating in 1902.

He began at Beyer, Peacock as a pupil in the Drawing Office and in 1904 joined the Midland Railway at Derby Locomotive Works, becoming an 'Improver' at Derby shed, then in 1905 moving to Woolwich Arsenal. In 1906, he joined the North Eastern Railway as Assistant to the District Locomotive Superintendent at Hull Dairycoates. In 1909, he moved on to a similar position at Gateshead and went with Vincent Raven on a study tour to the USA in 1910. After Nigel Gresley's promotion to Locomotive Supt. of the GNR in 1911, Thompson became Carriage & Wagon Supt. at Doncaster Works.

He married Vincent Raven's younger daughter, Guendolen, in 1913 and in 1914 transferred to Peterborough as Assistant District Locomotive Superintendent. Raven was appointed as Superintendent of the Royal Ordnance Factory at Woolwich in 1915 and Thompson joined him in 1916 in the Movements branch of the Directorate of Transportation in France, where he was awarded the rank of Lt. Colonel and subsequently the OBE.

After the war, he became Superintendent of the GNR Carriage & Wagon Works and introduced the concept of conveyor belt construction after the Henry Ford example. Gresley moved him to Stratford in 1923 as Assistant Mechanical Engineer, and he occupied a flat in Baker Street. He visited Paddington on many occasions and was impressed with the sharp exhaust of the GW 4-6-0s. He learned of that company's long valve travel arrangements and economies and was influential in obtaining authority for the rebuilding of the GE B12s in 1933. However, his staff management was poor and his relationship with Gresley was not good. Gresley had criticised him on occasions in front of colleagues and staff and he resented this bitterly.

In 1933 Thompson was appointed as Mechanical Engineer for the North East Division at Darlington and was there five years, his wife dying during this period. In March 1938 he moved to Doncaster as Mechanical Engineer there and in April 1941 Gresley died suddenly. Although it seems that the LNER management was grooming J.F. Harrison as Gresley's successor, it was felt that he was still too young and Thompson, already sixty years of age, was appointed instead.

For years, Thompson had felt neglected and left outside of Gresley's core team and seized this opportunity he had been given with just five years left before retirement. The fact that this coincided with the Second World War was for him both a problem and opportunity, for the LNER needed engines that were simple and robust to stand up to intensive use and lack of maintenance. Thompson had been critical of Gresley's 3-cylinder conjugated valve gear arrangement and lack of standardisation in the way both the GWR and subsequently the LMS under Stanier had developed for their respective systems. Thompson drew up a similar list of requirements for the LNER with ten standard types although only the B1, K1 2-6-0 and L1 2-6-4T were completely new designs, and persuaded the LNER Board to authorise this, highlighting the failures of the Gresley 3-cylinder engines in traffic, especially during the first years of the war. The B1 4-6-0 concept was a more conventional replacement for Gresley's V4 2-6-2 of which only two had been built. That and the K1 were successful and the rebuilding of some of the Robinson O4 2-8-0s with the 100A boiler as O1s also was useful. However, the other standard requirements on his list were basically achieved by rebuilding some of Gresley's three cylinder engines, the A1/1 *Great Northern*, the P2 2-8-2s as A2/2s, the other A2 variants, the B17/B2 conversion, the K3/K5 and one two inside cylinder rebuild of Gresley's D49/2, which is described in this book.

Gresley's loyal former staff were unhappy in this regime and his difficult manner had become unpredictable varying from charming, especially with women, to being autocratic and ruthless. He was a good chooser of assistants however and was efficient and cut costs at a time when this was a priority for the Company and the war effort. He retired in 1946.

Chapter 2
THE NORTH EASTERN 4-4-0s

The chronological order of the construction of the North Easrtern 4-4-0s was not followed by the LNER in classifying the different groups of engines, and whilst I deal with the classes in the LNER order for easy reference, for clarity I list below the classes in the order in which they were designed, built and rebuilt.

LNER class	NER class	Wheel Diameter	Built	Designer	Rebuilt
-	'38'	6' 8¼"	1884 – 1885	McDonnell	
D22	D, F, F1	6' 8"	1886 – 1891	T.W. Worsdell	1903-1908 Wilson Worsdell
D23	G	6' 1¼"	1887 – 1888	T.W. Worsdell	
D17/1	M	7' 1¼"	1892 – 1894	Wilson Worsdell	
D19	3CC	7' 1¼"	1893	Wilson Worsdell	
D17/2	Q	7' 1¼"	1896 – 1897	Wilson Worsdell	
D18	Q1	7' 7¼"	1896	Wilson Worsdell	
D20	R	6' 10"	1899 – 1907	Wilson Worsdell	
D21	R1	6' 10"	1908 – 1909	Wilson Worsdell	
D24	H & B R 'J'	6' 6"	1910	Kitson	

The NER class 38, McDonnell

Twenty-eight 4-4-0 passenger engines were designed and built by Alexander McDonnell in 1884 and 1885 at a cost of £3,000 each. Sixteen were constructed at Gateshead in 1884 and were given replacement numbers – 38, 112, 126, 158, 180, 186, 231, 234, 281, 385, 426, 500, 576, 664, 1318 and 1331. They became known as the '38' class after the first example to be built. A further twelve were ordered from R & W Hawthorn & Co. at £2,951 each and were delivered in 1884 and 1885 and were numbered in a new series, 1492–1503. Their key dimensions were:

Cylinders (2 inside): 17 x 24in
Coupled wheel diameter: 6ft 8¼in
Bogie wheel diameter: 3ft 1¼in
Stephenson motion with slide valves
Boiler pressure: 140lbs psi
Heating surface: 1,071.3sq ft
Grate area: 16.95sq ft
Axleweight: 14 tons 12 cwt
Weight (Engine): 37 tons 10 cwt
(Tender): 27 tons 14 cwt
(Total): 65 tons 4 cwt
Water capacity: 2,500 gallons
Coal capacity: 4 tons
Tractive effort: 11,760lbs

The clear intention had been to produce locomotives for the North Eastern Railway's main expresses to replace Edward Fletcher's 1872 class '901' 2-4-0s, but it became obvious very quickly that these engines were underpowered for such main line duties and not of equal standing with 4-4-0s now being constructed for other companies such as the Midland Railway. Another source of complaint was that the Hawthorn constructed locomotives needed frame strengthening and coupled wheel spring modification and this and problems in the reorganisation of the Works that McDonnell undertook caused him to tender his resignation in 1884.

They were tested on the main line in the summer of 1885 in comparison with the Fletcher '901' class and the Tennant '1463' class and were shown to burn 9 per cent more coal than the Fletcher 2-4-0 and 21 per cent more than the Tennant engine. As a result, the class '38s' were soon relegated to secondary lines where they performed satisfactorily for many years – the majority in the York, Scarborough, Whitby area with some working between Newcastle and Carlisle.

They were fitted with Westinghouse air brakes and twelve new short wheel base tenders were constructed as many of the secondary routes on which they were employed did not have sufficiently long turntables for a 4-4-0. They were reboilered with a standard Worsdell boiler between 1895 and 1900, giving more generous water spaces but a slight reduction in heating surfaces and grate area.

All, apart from one, No.281 of the 1884 Gateshead build, were withdrawn before the Grouping and 281 did not last long enough to

Above: **McDonnell class '38'** 1500 built by Hawthorn & Co in 1885 and seen here at Derby, Midland Johnson 4-4-0 1524 in the background. 1500 was reboilered later in 1896. (MLS Collection)

Below: **McDonnell class '38'** 426 built at Gateshead in 1884 and reboilered in 1896, seen here at Gateshead, c1897. 426 was withdrawn in 1915. (V.J. Bradley/MLS Collection)

McDonnell class '38' 158, built in 1884, reboilered in 1899 and seen here c1910. (Real Photographs/MLS Collection)

Below left: McDonnell class '38' 426 on shed alongside Fletcher '901' class 2-4-0 No. 366, c1912. (MLS Collection)

Below right: Tennant '1463' 2-4-0, No.1506, built at Gateshead in 1885 after McDonnell's resignation, reboilered in 1892 and withdrawn in 1926, seen here c1900. (F. Moore/MLS Collection)

be allocated an LNER class identity, being withdrawn in February 1923. After years based at Whitby, 281 spent its final months at York where it was spare for most of the time, though it was used on occasions for inspection saloon work.

The NER M & Q, LNER D17/1 & D17/2, Wilson Worsdell

Wilson Worsdell's older brother had built 4-2-2s and compound 4-4-0s before his retirement through ill health in 1890, though the latter's construction was not completed until April 1891 (see LNER D22 later). Wilson had put the design of a 4-4-0 to the Board in 1892, but the initial design failed to get approval. However, in April the Board gave instructions for twenty simple and one compound 4-4-0 to be constructed and the twenty were identified as class 'M1', later just simply 'M'. The twenty 'simple' class 'M1s' were built at Gateshead between 1892 and 1894 at a cost of £3,110 each and were numbered 1620 – 1639. Their key dimensions were:

Cylinders (2 inside):	19 x 26in
Coupled wheel diameter:	7ft 1¼in
Bogie wheel diameter:	3ft 7¼in

Stephenson motion with 8¾in piston valves*

Boiler pressure:	160lbs psi (increased to 175lbs psi later)
Heating surface:	1,341sq ft
Grate area:	19.6sq ft
Axleload:	18 tons 4 cwt
Weight (Engine):	50 tons 14 cwt
(Tender):	41 tons 4 cwt
(Total):	91 tons 18 cwt
Water capacity:	3,940 gallons
Coal capacity:	5 tons
Tractive effort:	14,974lb
Brakes:	Westinghouse and Vacuum

* As built 1620–1638 had outside flat valves for the inside cylinders, 1639 was built with piston valves. 1620–1638 were rebuilt with piston valves between 1903 and 1908. 1639 in a 1902 test recorded by the NER draughtsman, W.M. Smith, demonstrated its economic superiority with an average coal consumption of 29.55lbs per mile compared with the class average of 34.08lbs.

A further ten engines were built in 1896 and were identified as class 'Q' differing from the 'Ms' by having slightly larger cylinder diameter – 19½ x 26in and flat valve inside the frames and a cab with clerestory roofs. Boiler pressure for these engines was 175lbs psi from the start. 1872–1880 were fitted with water scoops and a reduced water capacity tender holding 3,375 gallons. All eventually had the larger 3,940 gallon tenders with water scoops. Their engine weight had been reduced to 49 tons 9 cwt. They were numbered 1871–1880. Another twenty class 'Q' were built in 1897 and were numbered 1901–1910 and 1921–1930. All the 'Qs' were later rebuilt with new cylinders and piston valves in line with the 'Ms'. Superheating of the whole class commenced in 1914 after conversion to piston valves

'M1' 1638 at York as built new in 1894 in NER passenger green livery with claret borders to splashers and tenders, introduced in 1886. Note the outside slide valves, and the air vent to the water scoop system in the tender. (Locomotive Publishing Co./MLS Collection)

had been completed, but not all had been superheated until 1929. The superheated boilers had revised dimensions of a reduced boiler pressure, 160lbs psi, heating surface of 1,097sq ft and grate area of 19.8sq ft. Engine weight increased to 52½ tons for the 'Ms' and 50 tons 2 cwt for the 'Qs'.

All were turned out in the NER passenger green livery and, like 4-4-0s of other constituent companies, started life after the Grouping in LNER lined apple green, only to be changed to black with red lining after 1928 as an economy measure. NER locomotives retained their numbers at the Grouping, the only company's engines to not require renumbering in 1923.

At the Grouping, the 'Ms' were reclassified as D17/1 and the 'Qs' as D17/2. The first to be withdrawn was 1628 in 1927 as the result of

'M' 1631 at Darlington, as built in 1893, after superheating and fitting with mechanical lubrication, c1920. (Real Photographs/MLS Collection)

D17/1 1629 at York, c1934. 1629 was the last survivor of the D17/1 class, one of two not withdrawn until 1945. (Real Photographs/MLS Collection)

D17/2 1871 at Hull Botanic Gardens, c1939. (Photomatic/MLS Collection)

D17/2 1874 at Hull Botanic Gardens a month before withdrawal, September 1938. Note cab part clerestory roof. A B16 4-6-0 is in the background. (W. Potter/MLS Collection)

One of the only two D17s to survive long enough to be renumbered in 1946, D17/2 2111, the former 1873. It is seen on York shed, 28 June 1947. It was withdrawn in February 1948 before being renumbered 62111. (N. Fields/MLS Collection)

being involved in an accident. Planned withdrawals commenced in 1931 and just two, 1621 and 1629, survived until the Second World War and were retained until its end, being withdrawn in July and September 1945. 1621 was set aside for preservation in the LNER's York Museum (see Chapter 7). Seven D17/2s were still in existence at the start of the war and five were withdrawn between 1943 and 1945. The LNER 1946 renumbering scheme allocated 2108–2114 to the surviving engines, but only two, 1873 as 2111 and 1902 as 2112, received these numbers. Both were allocated BR numbers, but they were withdrawn in February 1948 before repainting.

Operation

The 'M1s' were initially based at York, Newcastle and St Margaret's Edinburgh and worked the key East Coast expresses between York and Edinburgh with an engine change at Newcastle where the trains reversed before the building of the King Edward Bridge in 1906. They were famously involved in the 1895 'railway races' to the north, initially double-heading with a 'J' 4-2-2, but as the competitive spirit grew, the loads were reduced to six coaches, 105 tons, and were worked by the 'M1s' alone over both sections. In the final week of the 'competition', 1624 covered the 80.6 miles from York to Newcastle in 83 minutes 24 seconds and 1621 the 124.4 miles on to Edinburgh in 125 minutes 34 seconds. On the final night, 21/22 August 1895, 1621 ran from York to Newcastle in 78½ minutes, average speed 61.6mph, an alarming 74mph average over the last 8.66 miles from Chester-le-Street to the outskirts of Newcastle.

1620 more spectacularly covered the 124.4 miles onwards in just 113 minutes at an average speed of 66mph, with apparently scant regard to speed restrictions round curves at Morpeth, Alnmouth, Berwick and Portobello, the latter 20mph restriction round the reverse curves allegedly passed at 82mph! This epic performance earned 1620 a role at the 1925 Stockton & Darlington centenary procession and earmarked it for preservation, although in the end it was scrapped and 1621 was preserved (see Chapter 7).

I have traced just one run with an 'M' on a main line service, when its performance matched that of the later 'R' 4-4-0s of 1900, though with a light load. It was on the Darlington–York racing stretch and as the log was undated, I'm unsure if the run was before the advent of the 'Rs' or whether it was a substitute for one of them. I suspect the latter, for the acceleration of the 2.20pm Newcastle did not take place until 1902, so I estimate that this run was in the early 1902-1905 period before the full sixty of the 'Rs' had been put into service. A slow start and loss of time to Northallerton was almost recouped between Thirsk and Alne and the 'M' only missed the timing of the 'fastest train in the British Empire' by fifteen seconds.

	Darlington–York, c1902-1905			
	1626, 'M', 145 tons			
	2.20pm Newcastle - Bristol			
Miles	Location	Times	Speed	Gradients
0	Darlington	00.00		
2.6	Croft Spa	-	55½	
5.2	Eryholme	07.35	60	1/438 F, 1/391 R
10.5	Danby Wiske	12.30	68	1/650 F
14.2	Northallerton	15.55	67 2 L	L
17.6	Otterington	19.00	67	1/629 F
21.9	Thirsk	22.40	72½ 1 ¼ L	1/629 F, L
26.2	Sessay	-	68	L
28	Pilmoor	28.05	65	L
31	Raskelf	-	71	1/741 F
33	Alne	32.20	74 ¼ L	L
34.5	Tollerton	-	70	L
38.6	Beningbrough	37.15	68	L
42.5	Poppleton Jcn	40.45	67 ¾ L	
<u>44.1</u>	<u>York</u>	<u>43.15</u>	<u>¼ L</u>	

Two 'M1s', 1629 and an unidentified one leaving York with a heavy express for Newcastle and Edinburgh, c1898. (John Scott-Morgan Collection)

The pioneer and 'Races to the North' hero, 'M1' 1620, departing from York with a northbound express, c1898.
(John Scott-Morgan Collection)

The 'Q' series replaced the 'M1s' on the most prestigious workings after 1896 although the 'M1s' continued on main line work. 1908 hauling 320 tons between York and Darlington produced an average power output of 585 drawbar horsepower, but loads were normally well below this. Some dispersion took place in the middle of the first decade of the 1900s with 1626 and 1629 being stationed at Botanic Gardens for the Hull–Doncaster–Leeds services and 1623 at Blaydon for the Newcastle–Carlisle route. In 1907 the allocation of the 'M' fleet was:

Gateshead:	3
York:	5
Tweedmouth:	7
Darlington:	2

By 1920, the only 'Ms' allocated to main line sheds for semi-fast and stopping passenger services were 1621, 1625, 1631 and 1638 at Gateshead and 1624, 1627, 1632, 1634, 1636 and 1637 at Tweedmouth. Botanic Gardens had the majority – nine. The 'Qs' were displaced from the main East Coast expresses by the 'R' 4-4-0s from around 1900 and their allocation in 1908 was:

Gateshead:	13
York:	8
Heaton:	4
Leeds:	7
Tweedmouth:	1

The introduction of the 'V' and 'Z' Atlantics after 1910 together with the 'Rs' removed the 'Ms' and 'Qs' from main line work and Scarborough, West Hartlepool and Carlisle all received an allocation. Train safety on the NER at the turn of the twentieth century left something to be desired and these 'M' and 'Q' 4-4-0s were involved in a number of accidents. In July 1897, 'Q' 1869 had a brake failure (human error) when running 24 minutes late on the 12.20pm Edinburgh–King's Cross that it took over at Berwick. It ran into the back of a Tynemouth train at Newcastle Central at around 20mph. The same engine was involved in a collision after passing a signal at danger in January 1908 on a New Year's Day return excursion from Leeds to Saltburn, running into empty stock at Middlesbrough. In 1894 'M1' 1622 after getting a 2-4-0 as pilot due to running short of steam missed signals in the fog north of Northallerton and ran into the back of a mineral train with spectacular damage but fortunately no fatalities. In later year after the Grouping, engines of the two classes were involved in more serious accidents. In August 1926, D17/2 1929 smashed into a new road charabanc at a crossing at Naworth on the line to Carlisle with eight coach fatalities and that of the crossing-keeper. The following year, D17/1 1628 was departing from Hull Paragon when it collided head-on with a D22 arriving with a commuter train from Withernsea, with twelve fatalities. Both 1628 and the incoming D22 were badly damaged and scrapped. The accident was caused by a signalman's error.

At the Grouping the allocation of both classes were:

Classes:	M (D17/1)	Q (D17/2)
Gateshead:	5	3
Hull Botanic Gardens:	10	
Tweedmouth:	4	
Alnmouth:	1	
Carlisle:		8
Scarborough:		6
Neville Hill:		5
Starbeck (Harrogate):		1
West Hartlepool:		7

Six of Hull's D17/1s went to Bridlington in 1925 and two Tweedmouth and two Gateshead engines went to Hull and Selby. The first withdrawal took place in 1927, but 1628 had been damaged in a collision. By 1939, only 1621 at Alnmouth and 1629 at Bridlington remained. 1621 was withdrawn from Alnmouth in July 1945, but 1629 was transferred round a number of sheds during the war. It was withdrawn in September 1945.

There are just a few recorded runs with D17s in their latter years. 1621 had an evening semi-fast service on the main line north of Newcastle, and there were a couple of runs on the Leeds–York and Leeds–Selby–Bridlington routes, details below:

Newcastle–Chevington, 24.6.1930

1621, D17/1

6 chs, 130 tons

5.16pm Newcastle – Alnmouth

Miles	Location	Times	Speed	Gradient
0	Newcastle	00.00		
0.5	Manors	01.45		
1.6	Heaton	03.20		
2.4	Benton Bank	04.20	48	1/295 R
5	Forest Hall	07.35	48	1/200 R
5.9	Killingworth	08.45	53	
7.7	Annitsford	10.50	52	L
9.9	Cramlington	13.30	49	1/224 R
11.5	Plessey	15.20	53½	
13.9	Stannington	17.45	62	1/217 F
16.6	Morpeth	21.15		
0		00.00		
2	Pegswood	03.20		L
3.6	Longhirst	04.55	62	1/208 F
6.7	Widdrington	07.40	67½	
9	Chevington	10.25		

Miles	Location	Times	Speed		Gradients
		York–Leeds, 24.8.1936			
		1871 D17/2			
		11 chs, 317/340 tons			
		2.42pm York–Leeds (ex Scarborough)			
0	York	00.00		T	
2	Chaloners Whin	04.30			
3.75	Copmanthorpe	06.50	45		L
7.5	Bolton Percy	11.30	48½		
8.75	Ulleskelf	13.00			L
10.75	Church Fenton	15.05	57		1/867 F
15.25	Micklefield	22.30	42		1/133 R
18.25	Garforth	26.15	48		1/160 R, L
21.25	Cross Gates	30.00			
23	Osmanthorpe	31.25	66		1/162 F
24.75	Marsh Lane	33.35			
25.5	Leeds	35.45		5 ¾ L	

Time was lost with an overloaded return Scarborough holiday train, mainly due to the slow start to Church Fenton.

Neville Hill's D17/2 1908 departing from York passing Holgate excursion platform with a train for Leeds, c1925. (Real Photographs/ MLS Collection)

Leeds–Selby, 29.5.1937

1905 D17/2

6 chs, 160 tons

1.47pm Leeds–Bridlington

Miles	Location	Times	Speed	Gradients
0	Leeds	00.00		
0.8	Marsh Lane	02.15		
2.6	Osmanthorpe	05.35	32	1/153 R
4.4	Cross Gates	08.15	40	1/158 R
7.3	Garforth	12.25	42/ sigs	1/158 R, L
9.7	Micklefield	15.40	sigs	1/150 F
12.9	South Milford	19.15	53	1/136 F
16.5	Hambleton	22.45	62	L
18.3	Thorpe Gates	24.45		
20.7	Selby	28.30	(27 net)	

D17/1 1633 on the 4.52pm Leeds–Bridlington with five coaches completed the run to Selby in 25½ minutes but intermediate times are not recorded. It continued taking 20½ minutes to Market Weighton for the 17.2 miles, but unfortunately no record exists of the subsequent 1 in 100 climb to Enthorpe.

The D17/2s worked the Leeds–Stockton–West Hartlepool, Scarborough–York–Leeds, Scarborough–Hull and Newcastle–Carlisle routes, although surprisingly 1880, 1927 and 1929 were loaned in 1927 to the Great Eastern section because of a shortage of power there. Further cascades of power took place in the 1930s, when the new D49s displaced the D20s ('Rs') which in turn the displaced the D17s from their secondary duties, leading

D17/2 1903 with a Scarborough–Leeds express near Kirkham Abbey, c1926.
(MLS Collection)

D17/2 1923 departing from York with a Scarborough–Leeds express, c1930. (F. Moore/MLS Collection)

Below left: **D17/2 1901** at Carlisle with a stopping train for Newcastle, June 1936. (W. Potter/MLS Collection)

Below right: **D17/2 2111** (the former 1873) leaving York with a local train for Pickering, York Minster just visible on the left of the photograph, 28 June 1947. (N. Fields/MLS Collection)

to more D17/2 withdrawals. At the end of the 1930s, the remaining D17/2s were at Carlisle, Haymarket and St Margaret's. One engine (1901) at Duns worked to St Boswell's and Berwick until June 1945. The others were at Botanic Gardens (1871 and 1873), Scarborough (1902) and Bridlington (1905). During the war, many movements took place, 1871 and 1905 finishing up at Alnmouth, 1873 (as 2111) and 1902 (as 2112) at York, the latter two being used mainly on York–Pickering trains and an occasional officers' inspection saloon special until their withdrawal in February 1948.

The NER Q1 (LNER D18)

Two 4-4-0s with large diameter coupled wheels were built at Gateshead simultaneously with the building of the 'Qs' at the time that the 'races to the north' were at their height. These two engines with their 7ft 7¼in driving wheels were built for speed, but by the time they were completed, the risks associated with the 'races' had been deemed too great and speeds had moderated. Although they were tested, there is no record of what maximum speeds these two engines were capable of. They were numbered 1869 and 1870 and their dimensions were:

Cylinders (2 inside): 20 x 26in
Coupled wheels: 7ft 7¼in
Bogie wheels: 3ft 7¼in

Stephenson motion with slides valves (1869) and piston valves (1870)
Boiler pressure: 175lbs psi
Heating surface: 1,216sq ft
Grate area: 20.7sq ft
Axleload: 18 tons 14 cwt
Weight (Engine): 50 tons 16 cwt
(Tender): 41 tons 2 cwt
(Total): 91 tons 18 cwt
Water capacity: 3,940 gallons
Coal capacity: 5 tons
Tractive effort: 16,953lbs

Their boilers were unique among NER engines and when due for renewal, were replaced by 'Q' engine saturated steam boilers in 1910/11. There was little difference in the heating surface and a slight reduction in the grate area to 19.8sq ft. 1870 was fitted with a superheated boiler in 1915 and 1869 in 1920 and its slide valves replaced by piston valves and its cylinders lined up to 19in. Boiler pressure was reduced to 160lbs psi by the Grouping and tractive effort had

1870 as built in 1896 with a test shelter.
(F. Moore/MLS Collection)

1870 with superheated 'Q' boiler, c1920. The large diameter wheels and covering huge splashers are especially impressive in this photograph.
(Real Photographs/ MLS Collection)

been to reduced to 13,990lbs for 1869 and 15,500lbs for 1870.

They were classified as D18s and their NER livery was replaced by LNER lined green after the Grouping and 1870 was still in that livery when withdrawn in October 1930. 1869 was repainted black with red lining in 1929, but it was withdrawn at the same time as 1870.

Operation

Both locomotives were initially allocated to Gateshead, able to run north or south to Edinburgh or York, though in fact they were mainly used to York where their high speed capability and lower tractive effort were more suited to the flatter route, especially the Darlington–York racing stretch where a fast Newcastle–Bristol express was scheduled to cover the 44¼ miles in 43 minutes at 61.7mph, the fastest booked train in the country at the time. 1869 was timed on one occasion to run this stretch southbound in 42 minutes 7 seconds with a load of eight vehicles, 200 tons, in heavy rain throughout, maximum speed 74mph at Tollerton. This is a step better than the 'M' log quoted in full earlier.

Both engines however spent most of their lives – twenty years – at Neville Hill, Leeds, working trains to York and Scarborough or Selby and Hull.

The M, later 3CC (LNER D19)

Thomas Worsdell, despite his retirement due to poor health, remained a consultant to the NER Board as his brother took command. Although the Board granted authorisation for Wilson Worsdell's twenty 'M' simple 2-cylinder 4-4-0s, Thomas clearly persuaded the Board to consider compounding and authority was given for one new locomotive to be constructed in 1893 as a 2-cylinder compound. 1619 was delivered from Gateshead Works in May 1893 at a cost of £3,267 (some £150 more than its twenty 'simple' sisters) and was classified 'M' and the first three 'simple' 'M' engines already in service were reclassified as 'M1s'. Its initial dimensions were:

Cylinders
(2 inside)   20 x 26in high pressure on LH side, 28 x 26in low pressure on RH side

Worsdell-von Borries compound system
Boiler pressure: 200lbs psi

The remaining key dimensions were identical to those of the simultaneously built 'M1s' (see page 17). On test working similar duties to the 'M1s' a saving of 10 per cent on fuel consumption was claimed. Despite this, no further 2-cylinder compounds were built for the NER. Wilson Worsdell's chief draughtsman was W.M. Smith and he persuaded Worsdell to rebuild 1619 as a 3-cylinder compound with a high pressure cylinder between the frames and two low pressure cylinders outside. The engine was rebuilt in 1898. A new boiler and larger grate was used. The compounding system devised by Smith was that subsequently used by Deeley for the first five prototype Midland Railway compounds. Its success

1870 at York with a train for Scarborough, April 1927. (MLS Collection)

The North Eastern 4-4-0s • 29

'3CC' 1619 after rebuilding in 1898 as a 3-cylinder compound on the Smith system, seen at Darlington, c1920. (Real Photographs/MLS Collection)

1619 after the Grouping but before repainting in LNER livery, at Leeds Neville Hill, 1926. (Real Photographs/MLS Collection)

With '3CC' still stencilled on the bufferbeam, D19 1619 stands in Hull Botanic Gardens shed, 26 June 1923. (W. Potter/MLS Collection)

was enough to use the system for two NER Atlantics in 1906, but Smith died shortly afterwards and compounding was not pursued by the NER, superheating taking priority in improving the performance and economy of the NER 4-4-0s. The dimensions of the rebuilt 1619 were:

Cylinders (1 inside high pressure):	19 x 26in
(2 outside low pressure):	20 x 24in
Stephenson motion with 8" piston valves for HP and slide valves for LP cylinders	
Boiler pressure:	220, reduced to 200, then 180lbs psi after 1913
Heating surface:	1,328,5sq ft reduced to 1,290sq ft on reboilering in 1913
Grate area:	23.4sq ft
Axleload:	20 tons
Weight (Engine):	53 tons 6 cwt
(Tender):	41 tons 4 cwt
(Total):	94 tons 10 cwt
Water capacity:	3,940 gallons
Coal capacity:	5 tons
Tractive effort:	14,504lbs

In 1914 it was reclassified as '3CC' and at the Grouping the LNER gave it the classification of D19.

Operation

Between 1893 and 1898 it was based at Gateshead for East Coast work to Edinburgh and York. After rebuilding in 1898, it remained on East Coast express work before moving the Leeds in the early

1900s and Hull Botanic Gardens in 1907. After the Grouping, it remained at Hull until 1926 when it was transferred to Bridlington, working the morning residential train over the 31 miles to Hull in 40 minutes on a regular basis. It also shared duties with D17/1 engines based there in the late 1920s. It was withdrawn in October 1930 at the same time as the other two non-standard 4-4-0s, the two D18s.

The NER 'R' (LNER D20)

Train loads on the East Coast main line continued to increase and the NER, responsible for the York–Edinburgh section, required something more powerful to match the GN Atlantics being used for the London–York portion. Wilson Worsdell therefore proposed a more powerful 4-4-0, developing the 'M1' class with a larger boiler and increased boiler pressure, Stephenson motion and outside admission piston valves.

Gateshead built ten new engines to this design, designated class 'R', and they were delivered in 1899, numbered 2011–2020. Their key dimensions were:

Cylinders (2 inside):	19 x 26in
Coupled wheel diameter:	6ft 10in
Bogie wheel diameter:	4ft 0in
Stephenson motion with 8¾in piston valves	
Boiler pressure:	200lbs (later reduced to 175lbs)
Heating surface:	1,527sq ft, 1,413sq ft (1906), 1,364sq ft (1907) 1,318.7sq ft after Grouping.
Grate area:	20sq ft
Axleload:	19 tons 13 cwt

Weight (Engine):	51 tons 14 cwt (54 tons 2 cwt for 1906/7 builds)
(Tender):	37 tons 18 cwt
(Total):	89 tons 12 cwt
Water capacity:	3,537 gallons
Coal capacity:	5 tons
Tractive effort:	19,459lbs (17,026lbs after boiler pressure reduced)

The locomotives were immediately successful and were multiplied quickly, a further twenty being built in 1900-1901, numbered 2021-2030 and 2101–2110. Ten more were built in 1906 and were given spare numbers: 476, 592, 707, 708, 711-713, 723-725. Then in 1907 another twenty: 1026, 1042, 1147, 1206, 1209, 1217, 1232, 1234, 1236, 1260 as first batch of ten and 1051, 1078, 1184, 1207, 1210, 1223, 1235, 1258, 1665 and 1672 as the second. All were fitted with both Westinghouse air brakes and vacuum brakes.

Superheating commenced in 1912 and all had been superheated by the Grouping apart from 1234 and 2025 which were superheated in 1929 and 1925 respectively. Some were fitted with Schmidt superheaters with 1,248sq ft of heating surface, while the Robinson superheater engines varied from 1,282 to 1,369sq ft of heating surface. The class was designated as D20 at the Grouping, their numbers remaining as the NER system. Like other 4-4-0s, they initially changed their NER green livery for LNER lined apple green, the black with red lining after 1928, plain black during the war years. Surprisingly, none of the fifty got the BR mixed traffic

Above: **The prototype** 2011 built in August 1899 and not withdrawn until February 1951, seen at York, c1903. (F. Moore/MLS Collection)

Below: **1026 built** in February 1907 and seen here in original condition, c1912. (MLS Collection)

lined black livery, remaining plain black with the BR lion and wheel emblem, though many had their brass beading over the splashers burnished – or at least kept clean – to offset the drab appearance.

A rebuilding was made in 1936 by Thompson (without consulting Gresley which drew a sharp public rebuke), 2020 being rebuilt at Darlington with larger (10in) diameter piston valves and long travel valve gear, after Thompson's experience at Stratford rebuilding the B12s, and was classified as a D20/2. Its Westinghouse air brake system was removed. After Gresley's death, Thompson intended the rebuilding of the remainder whilst maintaining their outline appearance. In the event, the war intervened, and no priority was to be given to his intention.

1051, built in June 1907, in LNER lined green livery, c1925. (MLS Collection)

1234, in LNER plain black but retaining its brass splasher beading, one of the first two to be withdrawn, seen at Hull Botanic Gardens five years earlier, in September 1938. (W. Potter/MLS Collection)

The North Eastern 4-4-0s • 33

707 built in 1906 at York with an Ivatt Atlantic, (Photomatic/MLS Collection)

The Thompson rebuild, 2020, with long travel gear, changed cab design and running plate clear of the coupled wheels, at York, May 1937. (W. Potter/MLS Collection)

The former 1235 built in 1907, renumbered 2392 in 1946, at Leeds Starbeck, 3 October 1948. (MLS Collection)

Below left: **The former** 2110, now E2369, seen at York in 1949. Note the new shape smokebox door. (MLS Collection)

Below right: **62383 ex-works** Darlington in May 1955 with rebuilt flat sided tender and large smokebox door. 62383 was one of the last withdrawals in May 1957. (T.K. Widd/MLS Collection)

62386 (formerly 1207) as seen in the latter days of the D20s, with large smokebox door, flat-sided tender and in run down condition, 1 January 1951.
(MLS Collection)

The rebuilt D20/2, 2020, renumbered 62349, at Heaton, 25 June 1950.
(H.C. Casserley/MLS Collection)

Three further engines were rebuilt by Thompson in 1942 and Peppercorn in 1948, (2101, 592 and 712, the latter then 62375) but no more as by this time it was already too late in their lives. Engine weight increased to 55 tons 9 cwt.

Tenders were sometimes exchanged with the 3,940 gallon tenders that the D17s had and after nationalisation ten D20s were equipped with tenders rebuilt with flat sided bodies similar to new LNER standard tenders. They held 3,900 gallons and 6 tons 5 cwt of coal. Sometime earlier – from around 1937 – water scoops were removed as the engines had no long distance working warranting their retention.

1147 and 1234 were withdrawn in 1943 but the remaining fifty-eight locomotives were renumbered 2340–2397 in 1946, and fifty were passed into British Railways

A graveyard for D20s, left to right, 62362, 62376 and 62366 in the scrap yard at Darlington North Road, 8 April 1951. (MLS Collection)

stock in January 1948 and became 62340–62397. Withdrawals took place steadily between 1951 and 1956 with just seven left at the beginning of 1957, all based at Alnmouth – 62375, 62381, 62383, 62387, 62395 and 62396. The four D20/2s were withdrawn in 1954 (62371), 1956 (62349, 62360) and May 1957 (62375). The last to be withdrawn were 62381, 62395 and 62396 in November.

Operation
The introduction of the 'Rs' had a significant impact on the North Eastern's express services. By 1901, thirty were in service and the pioneer, 2011, had already amassed over 284,000 miles in traffic before its first general overhaul. Other members of the first batch of ten were running mileages of 160,000 and more before their first Works visit. The initial allocation of the 'Rs' was to York, Gateshead and Haymarket, where they were used on the York–Newcastle and Newcastle–Edinburgh expresses, displacing the 'M1s' and 'Qs' on the most important and fastest services. Despite the addition of ten 'V' Atlantics in 1903, thirty more 'Rs' were built in 1906 and 1907 and logs of their performance frequently appeared in the regular 'Locomotive Performance and Practice' articles of Cecil J Allen in the *Railway Magazine*. Although most of the express services were not unduly fast and weighed between 200 and 300 tons, there was one train booked at over 60mph start-to-stop, that the NER publicised as the 'Fastest Train in the British Empire', the 12.20pm Newcastle to Bristol that from July 1902 was booked to cover the 44.1 miles from Darlington to York in 43 minutes. In the reverse direction, the 9.13pm from York was allowed just 44 minutes until a minute was added in 1909 as the train load had increased. The most spectacular performance recorded was with 1672, the last 'R' built in September 1907, which completed the Darlington–York section in 39 minutes 34 seconds with a maximum speed of 79mph, and averaging over 70mph for most of the journey.

Darlington–York, 1907
12.20pm Newcastle–Bristol (1.09pm ex Darlington)

		1672, 152/170 tons			1207, 152/165 tons			1207, 135 tons			
Miles	Location	Times	Speed		Times	Speed		Times	Speed		Gradient
0	Darlington	00.00			00.00			00.00			
2.6	Croft Spa	03.54			04.32			-			
5.2	Eryholme	06.14	72		07.13	69		07.10	60/68		1/438 F, 1/391 R
10.5	Danby Wiske	10.33	76		11.42	72 ½		11.35	76½		1/650 F
14.2	Northallerton	13.46	73/70		15.01	67		14.45	72		L
17.6	Otterington	16.36	71		18.01	68		17.40	70		1/629 F
21.9	Thirsk	20.04	79	1½ E	21.32	75	T	21.05	77½	½ E	1/629 F, L
26.2	Sessay	23.33	76		25.16	71		-			L
28	Pilmoor	25.09	74		27.00	68		26.10	74		L
31	Raskelf	27.23	73		29.20	74		-			1/741 F
33	Alne	29.09	72	2¾ E	31.12	68	¾ E	30.05	78	2 E	L
34.5	Tollerton	30.20	75		32.27	72		-	77		L
38.6	Beningbrough	33.45	74		36.08	70		34.30	75		L
42.5	Poppleton Jcn	37.00	75/72	3 E	39.32		½ E	40.00	pws severe		L
44.1	York	39.34		3½ E	42.33		½ E	42.50 (40 net)		¼ E	

'R' 1207 departing York for Leeds on the 12.20pm Newcastle–Bristol, just after completing its 43 minute 61mph average Darlington–York section, the fastest in Britain at the time, c1910. This engine was recorded at covering four miles at an average of 80mph on this train in the face of a strong south west gale.
(MLS Collection)

1147 heads a Newcastle–Edinburgh express over Lucker water troughs between Chathill and Beal, c1912. 1147 was in fact the first 'R' to be withdrawn but not until 1943. (F. Moore/MLS Collection)

A sample of other runs from Cecil J's articles are tabled below. The first table is another set of runs on the 12.20pm Newcastle a couple of years later and a log of trains in the northbound direction on the 9.13pm from York.

		Darlington–York, 1909						
		12.20pm Newcastle–Bristol (1.18pm ex Darlington)						
		1207		1672		1209		
		5 chs, 130 tons		6 chs, 145 tons		5 chs, 130 tons		
Miles	Location	Times	Speed	Times	Speed	Times	Speed	Gradients
0	Darlington	00.00		00.00		00.00		
2.6	Croft Spa	04.18		04.32		04.14		1/438 F
5.25	Eryholme Jn	06.58		08.02	pws	06.59	56½	1/391 R
7	Cowton	08.33	66	10.02		08.44	64	1/490 F
10.4	Danby Wiske	11.33	69	13.07	70	11.54	66	1/650 F
14.2	Northallerton	14.48	71	16.17	74	15.19	69	L
17.6	Otterington	17.48	70	19.17	70	18.24	67	1/629 F
22	Thirsk	21.13	78	22.42	78	22.04	75	L
26.2	Sessay	24.48	71	26.07	74	25.49	67	L
28	Pilmoor	26.23	70	27.42	72	27.29	65	L
31	Raskelf	28.38	75	29.52	80	29.49	77	1/741 F
33	Alne	30.38	72	31.32	75	31.34	68	L
34.5	Tollerton	31.48	75	32.47	72	32.49	74	L
38.6	Beningbrough	35.38	73	36.12	72	36.34	72	L
44.1	York	42.00	1 E	42.13	(net 41) ¾ E	42.29	½ E	

York–Darlington, 1907-09
9.13pm York–Newcastle

Miles	Location	1907 2102, 200 tons Times	Speed	1909 1147, 200 tons Times	Speed	1909 1042, 200 tons Times	Speed	Gradients
0	York	00.00		00.00		00.00		
5.5	Beningbrough	07.55		07.41		08.05		L
9.6	Tollerton	11.40	66	11.41	62	12.05	64	L
11.1	Alne	12.55	72	13.06	65	13.30	66/70	L
13.1	Raskelf	14.40	69	15.06	60	15.20	69	1/666 R
16.1	Pilmoor	17.10	69	17.46	62	17.50	69	L
17.9	Sessay	18.50	69	19.41	63	19.40	69	L
22.1	Thirsk	22.35	67	23.36	64	23.20	67	L
26.5	Otterington	26.30	67½	27.36	67	27.10	69	1/629 R
29.9	Northallerton	29.30	68	31.06	59	30.20	64	1/671 R
33.7	Danby Wiske	32.50	69	34.46	64	33.40	67	L
37.1	Cowton	35.45	70	37.56	66	36.50	65	1/650 R
39	Eryholme	37.15	69	39.41	64	38.25	pws	1/490 R
41.5	Croft Spa	39.30	72	42.01	67	41.10		1/391 F
44.1	Darlington	42.50	1½ E	45.07	T	44.38	(43 ¾ net) ½ E	

Although the 'Rs' were best known for speedy haulage of comparatively light trains, Rous-Marten recorded 2011 on a 15-coach 340 ton train running the 36.5 miles from Newcastle to Darlington in 44 minutes 42 seconds, and the 44.1 mile Darlington–York stretch in 48½ minutes in 1900. Professor Tuplin, in his book *North Eastern Steam*, calculated that 2013, still unsuperheated, produced an average drawbar horsepower figure of 725 when hauling 375 tons between Poppleton Junction and Eryholme, 35 hp per sq ft of grate area, a figure superior to other NER 4-4-0s. The 44.1 miles on this northbound run on the Down 'Flying Scotsman' were covered in just under 49 minutes with an average speed of 58.2mph between Alne and Croft Spa, with a maximum of 62mph. 2028 on the Up 'Flying Scotsman' with 365 tons including the dynamometer car, achieved a time of 46 minutes 56 seconds with a maximum of 65mph at Thirsk and 64 at Raskelf but had to work hard, with 40 per cent cut-off and regulator ¾ open all the way.

The 'Rs' as the most numerous of the NER express passenger engines seem to have had their fair share of involvement in train accidents. In November 1906, 592 working the 2.20pm King's Cross ran into derailed wagons foul of the Down line at East Linton north of the border, with considerable damage but luckily no fatalities. In March 1907, 725 was derailed on buckled track on the 8.35am Liverpool–Newcastle at Felling just short of Newcastle with two fatalities. Then, later in 1915, 'R' 1217 was in collision with 'M1' 1622 at Micklefield after passing a signal at danger. Another collision occurred at Selby in 1921 as 'R' 2020 pulled away from the platform against signals and was struck by a passing train.

A run from Edinburgh to Newcastle with 592 and 220 tons was described by Cecil J. Allen in a 1923 *Railway Magazine* run with an even heavier load was recorded with 1260 and a train of 300 tons including a dynamometer car. I show the logs on below:

The latter test run was compared with a 'V' Atlantic and comments were made that despite loss of time on the uphill sections, the 'R' was a better hill climber, but the Atlantic was faster downhill.

The only NER locomotives allocated to Haymarket until the First World War were the 'Rs' where they were working to schedules of 78-82 minutes or so for the 67 miles between Newcastle and Berwick with trains of up to 300 tons. The first impact on their coverage of the main East Coast services north of Newcastle was the introduction of Raven's 3-cylinder 'Z' Atlantics, followed in the 1920s after the Grouping by the introduction of Gresley's Pacifics and their proliferation to include Haymarket and Gateshead allocations.

One of the services worked by Starbeck 'Rs' was the extension from Harrogate to Darlington of expresses from King's Cross routed via Leeds. The 11.20am, loaded to 315 tons, arrived very late and 476 backed on with a driver determined to recover what time he could.

Edinburgh–Newcastle, c1912

592, 220 tons   **1260, 300 tons**

5.15pm Edinburgh–Leeds

Miles	Location	Times	Speed		Times	Speed		Gradients
0	Edinburgh Waverley	00.00			00.00			
3	Portobello	04.40		1¼ E	05.10			1/78, 1/300 F
6.5	Inveresk	08.25		2½ E	-			
13.3	Longniddry	15.10	60	3¾ E	-			
17.8	Drem	19.45	66	4¼ E	22.50		1¼ E	1/300 F
23.5	East Linton	25.10	69/63		-			1/250 R
29.3	Dunbar	30.30	69	5½ E	35.05		1 E	1/360 F
36.5	Cockburnspath	38.55	51/42		43.55			1/96 R
41.3	Grantshouse	48.15	26	3¾ E	54..40	22	2¾ L	1/96 R (slipping)
46.3	Reston Junction	53.55	72	4 E	60.20	69	2¼ L	1/200 F
52	Burnmouth	60.00	easy		65.35			L, 1/190 F
		-			sigs stand			
<u>57.5</u>	<u>Berwick-on-Tweed</u>	<u>67.20</u>		4¾ E	<u>72.35</u> (71 net)		1¾ L	
0		00.00		T	00.00		T	
1.2	Tweedmouth	03.15			03.00			
8.3	Beal	14.15	sigs		-			1/190 F, L
15.3	Belford	21.35	58	3½ L	17.40		¼ E	1/208 R, L
20.9	Chathill	27.10	66		-			L
	Christon Bank	-	50		27.05			1/150 R
32.1	Alnmouth	38.20	70½	¾ E	36.20		¾ E	1/170 F
38.4	Acklington	45.00	55*		-			
50.3	Morpeth	58.35	pws	1½ E	57.00		T	
<u>67.9</u>	<u>Newcastle-upon-Tyne</u>	<u>80.30</u>		1½ E	<u>78.55</u>		T	

Harrogate–Darlington, c1912
476, 315 tons
11.20am King's Cross (3.32pm ex Harrogate)

Miles	Location	Times	Speed		Gradients
0	Harrogate	00.00		40 L	
3.4	Nidd Bridge	04.35			1/66 F, 1/255 F
6.6	Wormald Green	07.45	60		1/357 F
11.5	Ripon	11.30	82	39½ L	1/133 F
17.2	Sinderby	16.30	68		1/294 R, L
24.2	Cordio Junction	23.20	62/ 40*		1/280 R
25.2	Northallerton	24.50	25*	34¾ L	
29	Danby Wiske	29.50		L	
34.3	Eryholme	35.35	55/60		1/490 R
	Croft Spa	-	72		1/391 F
39.4	Darlington	40.55		35 L	

In 1923 on the first day of the Grouping the allocation of the D20s as they were classified by the LNER was:

Gateshead:	14
York:	15
Darlington:	7
Tweedmouth:	8
Heaton:	3
Leeds Neville Hill:	7
Hull Botanic Gardens:	10
Blaydon:	1

Little changed during the 1920s, the allocation remaining relatively static, Gateshead giving one to Darlington and Neville Hill

D20 2021, built in 1900, leaving York for Leeds with a mail train for Birmingham and the West of England, c1925. (Real Photographs/MLS Collection)

1260 on a Scarborough–Leeds express, near Kirkham Abbey, c1930. (MLS Collection)

three to Scarborough, with York transferring one to Scarborough and one to Starbeck (Harrogate). They worked secondary services over the main line and regular trains between Leeds and Hull, Leeds and Scarborough, Leeds and Northallerton/West Hartlepool and Newcastle–Carlisle. Some of these secondary services could be timed quite tightly such as a 75 minute schedule for Leeds to Scarborough, on which the pioneer 2011 was timed in the summer of 1929 passing York in 27 minutes and then just 45 minutes for the 42 miles onto Scarborough knocking three minutes off the schedule. The 8.25am Scarborough–Hull was timed to cover the 31 miles from Bridlington to Hull in 37 minutes and this turn was worked by a D20 up to the beginning of the Second World War. The next significant impact on the allocation and use of the D20s was the introduction of Gresley's D49 4-4-0s between 1927 and 1935. The allocation at the end of 1935 was:

		Change from 1923
Gateshead:	3	-11
York:	10	-5
Darlington:	5	-2
Tweedmouth:	6	-2
Heaton:	3	No change
Leeds Neville Hill:	3	-4
Hull Botanic Gardens:	10	+5
Blaydon:	1	No change
Selby:	2	+2
Starbeck:	8	+8
West Hartlepool:	4	+4
Stockton:	1	+1
Scarborough:	4	+4

As the 1930s progressed and the Gresley engines (A3s, A4s and D49s) dominated work in the North East, and the earlier 4-4-0s were withdrawn, the D20s found themselves covering the secondary and local services that particularly had been the domain of the D17s. Alnmouth had acquired four for stopping services on the main line and Selby and Starbeck had significant increases in numbers at the expense of York in particular. A few sample logs taken from the Rail Performance Society's archives show typical rather than exceptional performances:

Leeds–York, 1931
708, 5 chs, 125 tons
2.45pm Leeds–Scarborough
18.7.1931

Miles	Location	Times	Speed	Gradient
0	Leeds	00.00	T	
0.8	Marsh Lane	01.40		
2.7	Osmanthorpe	04.15	44	1/162 R
4.4	Cross Gates	06.20	46	1/160 R
7.3	Garforth	10.15	44½/52	1/160 R/ L
9.7	Micklefield	13.00	60	1/150 F
14.8	Church Fenton	18.05	50*	
16.5	Ulleskelf	20.05	57	L
17.9	Bolton Percy	21.25	55	L
21.8	Copmanthorpe	25.25	58½	L
23.6	Chaloner's Whin	27.20	56½ / sigs	
25.6	York	34.45	(30½ net)	3¾ L

York–Scarborough, 1936
724, 7chs, 186 tons
2.40pm York–Scarborough
4.4.1936

Miles	Location	Times	Speed	Gradient
0	York	00.00	T	
4.2	Haxby	07.25		L
6.6	Strensall	10.00	55	L
9.2	Flaxton	13.00	52	1/310 R
11.5	Barton Hill	15.20	61	1/272 F
15	Kirkham Abbey	19.30	45*	
15.8	Castle Howard	20.35	45*	
18.4	Huttons Ambo	23.30	54	
21.2	Malton	27.40		¼ E
0		00.00	T	
4.4	Rillington	06.30		L
6.4	Knapton	08.35	58	L
8.3	Heslerton	10.30	pws	
11.7	Weaverthorpe	15.28		L
13.3	Ganton	17.40		
18	Seamer	22.50	54	1/255 R
20.9	Scarborough	27.05	(25 net)	2 L

Leeds–Harrogate, 1936

1236, 7 chs, 167 tons

1.48pm Leeds–Harrogate

Miles	Location	Times	Speed	Gradient
0	Leeds	00.00	T	
0.7	Holbeck	02.55	sigs stop	
2	Cardigan Road	11.00	33	1/96 R
3	Headingley	12.50	pws	1/100 R
5.8	Horsforth	17.45	33 ½	1/100 R
9.2	Arthington	22.30	45*	1/94 F
11.8	Weeton	25.20	50	1/195 R
13.2	Rigton	27.10	50	1/195 R
15	Pannal	29.10	52½	1/257 F
16.2	Crimple Junction	30.57	41	1/91 R
<u>18.2</u>	<u>Harrogate</u>	<u>35.00</u>	(28¼ net) <u>3 L</u>	

Harrogate–Leeds, 1937

1236, 7 chs, 140 tons   1206, 6 chs, 125 tons

8.10pm Harrogate   8.10pm Harrogate

17.2.1937   3.5.1937

Miles	Location	Times	Speed	Times	Speed	Gradient
0	Harrogate	00.00	T	00.00	T	
2	Crimple Junction	04.00		03.50		1/91 F
3.2	Pannal	06.10		05.40	46	1/257 R
5	Rigton	-	52	08.10	58	1/195 F
6.5	Weeton	09.55	62	09.50		1/195 F
9	Arthington	12.25	60	12.45	52	1/94 R
12.5	Horsforth	17.55	38/30	18.30	36/28	1/94 R
15.2	Headingley	20.45	60	21.28	59	1/100 F
17.5	Holbeck	<u>24.30</u>	1½ E	25.10	sigs	
<u>18.2</u>	<u>Leeds</u>			<u>34.55</u>	(31 net) <u>3 L</u>	

During the war years, a few returned to Gateshead and Darlington mainly to act as pilots to heavily loaded wartime passenger services. One D20 sustained war damage. 2108 on a Pickering–York train fell into a bomb crater in the line at Coxwold. The first D20 to be withdrawn was 1147 in 1943 when it was not considered easily repairable, but the majority carried on with fifty of the original sixty still operating from the following sheds at nationalisation on 1 January 1948:

		Change since 1935
Starbeck:	10	+2
Selby:	9	+7
Alnmouth:	7	+7
Hull Botanic Gardens:	5	-5
Tweedmouth:	3	-3
Duns:	1	+1
Blaydon:	3	+2
Northallerton:	3	+3
York:	1	-9
West Hartlepool:	3	-1
Stockton:	2	+1
Bridlington:	3	+3

The following depots had lost their allocation of D20s completely: Gateshead, Heaton, Darlington, Leeds Neville Hill and Scarborough, mostly losing out to D49s. After the war, holiday traffic to the East Coast resorts began to build again and the Selby and Bridlington D20s in particular found work on summer trains to Scarborough, Filey and Bridlington, often double-headed with another D20, D49 or LMR Ivatt 2-6-0. The Selby and Starbeck D20s found work on smartly timed local services in the York, Harrogate, Leeds and Doncaster area. The specific allocations in January 1951 were:

Selby:	62340, 62341, 62348, 62361, 62363, 62366, 62374, 62378, 62381, 62382, 62384, 62386, 62395
Starbeck:	62342, 62343, 62369, 62370, 62373, 62389, 62392, 62397
Tweedmouth:	62344
York:	62345
Northallerton:	62347, 62359, 62388, 62391
Alnmouth:	62349, 62351, 62352, 62354, 62358, 62360, 62362, 62371, 62380, 62387
Bridlington:	62353, 62355, 62375
West H'pool:	62372, 62379
Hull B.Gdns:	62383, 62396

I give below some examples of their post-war running with logs from the Railway Performance Society's archives.

Leeds–Selby, c1948

2395

6 chs, 215 tons

Miles	Location	Times	Speed	Gradients
0	Leeds	00.00		
0.8	Marsh Lane	02.27		
4.4	Cross Gates	08.20	38½/36½	1/162 R
7.2	Garforth	12.40	37½	1/160 R, L
9.6	Micklefield	16.50	sigs 28*	1/133 F
12.9	South Milford	21.30		1/136 F
		00.00		
3.7	Hambleton	05.36	53/sigs	1/136 F, L
8	Selby	11.46		

A filthy Starbeck D20, 2370, with a northbound stopping train at York, 28 June 1947. (G. Shuttleworth/MLS Collection)

Leeds–Selby, 1952
62386
7 chs, 202 tons
11.50am Leeds – Hull

Miles	Location	Times	Speed		Gradients
0	Leeds	00.00		1 L	
0.8	Marsh Lane	02.27	30		
2.7	Osmanthorpe	05.43	38½		1/153 R
4.4	Cross Gates	08.12	40½	1¼ L	1/158 R
7.3	Garforth	12.02	50		L
9.7	Micklefield	14.49	54/eased		1/150 F
12.9	South Milford	18.34		1 L	
0		00.00		1 L	
3.6	Hambleton	05.06	58		L
7.7	Selby	10.22		1¾ E	

As numbers dwindled, the active fleet at the beginning of 1955 was allocated as follows:

Selby: 62343, 62378, 62381, 62384, 62386, 62395
York: 62345
Blaydon: 62349
Alnmouth: 62355, 62360, 62375, 62383, 62396
Northallerton: 62359
West H'pool: 62372
Heaton: 62387
Neville Hill: 62397

A number of special railtours ran with D20 haulage in the 1950s. 62360 was involved in an SLS railtour in 1952 and there were

Thompson rebuild D20/2 62371 on shed at Carlisle London Road with B1 61238 *Leslie Runciman*, 30 August 1952. (MLS Collection)

Heaton's 62387 on the 11.48am Leeds–Hull express on the climb to Cross Gates, 6 August 1955. (MLS Collection)

three railtours in June 1957 using 62387, one an RCTS six-coach special which included a run from Leeds to York and back with 33mph minimum on the 1 in 160 at Cross Gates and a maximum of 59mph on the outward run and 32mph on the 1 in 133 to Micklefield and a maximum of 54 on the return. In their last year of operation, 1957, the following were all based at Alnmouth: 62375, 62381, 62383, 62387, 62395 and 62396. 62397 was withdrawn in February 1957 from Bridlington, transferred there in June 1956 from Neville Hill. The last three, 62381, 62395 and 62396, were withdrawn in November.

Selby's 62395 and 62381 at Bridlington on the summer Saturday 11.20am Scarborough–Liverpool holiday express, which they will work as far as Gascoigne Wood, August 1954, c1955. (J.W. Armstrong/MLS Collection)

Northallerton's 62347 at Hawes with the 4pm to Northallerton on the last day of the service, 24 April 1954. (MLS Collection)

Below left: **D20/2 62360** working a Manchester–Hull Stephenson Locomotive Society railtour at Cudworth, 24 August 1952. (A.C. Gilbert/MLS Collection)

Below right: **One of** the last six D20s active in 1957, Alnmouth's 62387, was selected for three railtours in June of that year. 62387 is at Leeds City with an RCTS tour for York on 22 June. (H.D. Bowtell/MLS Collection)

The NER 'R1' (LNER D21)

By the end of the twentieth century's first decade, train loadings were increasing substantially and despite the prowess of Wilson Worsdell's 'Rs', there was a need for a more powerful locomotive to avoid the necessity of double-heading. There were the ten 'V' Atlantics and the two '3CC' compounds but Worsdell decided to build on the success of the 'Rs' by designing an enlarged version of them, using the same cylinder and wheel diameter sizes as the earlier 4-4-0 but the enlarged boiler of the Atlantic. Ten locomotives of the new design were constructed in 1908 and 1909 and were numbered 1237–1246 and given the classification 'R1'. Their key dimensions were:

Cylinders
 (2 – inside): 19 x 26in
Coupled wheel
 diameter: 6ft 10in
Bogie wheel
 diameter: 3ft 7¼in

Stephenson motion with 10in piston valves
Boiler pressure: 225lbs psi (later reduced to 175lbs psi)
Heating surface: 1,737sq ft
Grate area: 27sq ft
Axleload: 20 tons 16 cwt
Weight (Engine): 59 tons
 (Tender): 46 tons 2 cwt
 (Total): 105 tons 2 cwt
Water capacity: 4,125 gallons
Coal capacity: 5 tons 10 cwt
Tractive effort: 17,026lbs (after boiler pressure reduction)

Notable was the axleload of 20¾ tons, the first express locomotive in the United Kingdom to exceed 20 tons.

The intention was for these locomotives to be allocated to the heaviest rather than fastest trains, but somehow they did not achieve the success of his 'Rs' and no more were built after these ten. For heavier loads thereafter, the NER resorted to the Atlantic design, despite the high adhesive weight of the 'R1s'. They were superheated between 1912 and 1915, with a reduction in total heating surface to 1,449.58sq ft including 390.7sq ft of superheater later calculated as 258sq ft. The smokebox was lengthened at the same time. The boiler pressure was significantly reduced then, initially to 160lbs psi, but raised to 175 before the Grouping. New boilers with Schmidt superheaters were built between 1920 and 1923. At the Grouping, the class was redesignated 'D21' and the NER green livery replaced by the LNER lined apple green until 1928, when in common with other 4-4-0s, they were painted black with red lining.

In 1941, Thompson considered rebuilding the 'D21s' with long travel valve gear on the lines of 'R' 2020 and 1237 was selected for rebuilding but wartime priorities meant that the plan was dropped and rebuilding with long travel valves was restricted to the handful of 'D20s'. Undoubtedly their extra power was useful during the war years, but after the decision was made not to rebuild them, they were withdrawn as they needed heavy repairs, 1239 being the first to go as early as December 1942. Only two survived the war, 1243 being withdrawn in June 1946 and 1245 in February 1946.

1238 as built in December 1908 in Works photographic grey livery at Darlington. (British Rail/MLS Collection)

Above left: **Superheated D21** 1239 with slightly extended smoke box in LNER lined green livery, c1925. (F. Moore/MLS Collection)

Above right: **1237, the** D21 selected for rebuilding with long travel valves, seen at Darlington awaiting the decision, c1941. (Colling Turner/MLS Collection)

Right: **1238 in** LNER black livery at Selby, September 1938. The Westinghouse air brake pump is visible – the 'R1s' were fitted with both air and vacuum brakes. (W. Potter/MLS Collection)

Operation

The 'R1s' were built for heavy haulage rather than speed – the 'V' Atlantics and the 'Rs' were appropriate for the lighter faster trains. A couple, 1242 and 1244 were allocated to Leeds Neville Hill for the morning Leeds–Glasgow service and the other eight were all based at York for York–Newcastle trains. They were also rostered to the heaviest trains between Newcastle and Edinburgh such as 'The Flying Scotsman', which would load to

350 tons or more but was allowed 98 minutes for the 80.1 miles from York to Newcastle and as much as 149 minutes for the 124.4 miles on to Edinburgh. On test they were found capable of sustaining 64mph on level track with 360 tons – enough to time NER schedules, but not outstanding compared with other express locomotives on other railways at the same period. A couple of logs of D21 performance on that service are available from the RPS archives and are tabled below:

York–Newcastle
1239, 342 tons
The Flying Scotsman

Miles	Location	Times	Speed	Gradients
0	York	00.00	2 ¾ late	
22.1	Thirsk	28.40		L
29.8	Northallerton	37.30	52.4 ave	1/629 R
44.1	Darlington	52.15	58.1 ave	L, 1/650 R, 1/391 F
57.1	Ferryhill	68.05	49.4 ave	L, 1/203 R
65.9	Durham	78.30	50.8 ave	1/200 F, 1/150 F, 1/150 R
80.1	Newcastle	95.35	¼ L	

A continuation of this run is logged below together with a run on a lighter train with a Neville Hill engine, probably on the morning Leeds–Glasgow service which had a slightly faster schedule of 138 minutes for the 124.4 miles.

Newcastle–Edinburgh

| | | 1237, 317 tons | | 1242, 200 tons | | | |
| | | *The Flying Scotsman* | | | | | |
Miles	Location	Times	Speed	Times	Speed		Gradients
0	Newcastle	00.00	¼ L	00.00		T	
16.6	Morpeth	22.00		23.50	pws	4¾ L	
34.8	Alnmouth	40.49	58.2 avr	42.50	57.5 ave	5¼ L	1.200 F, undulating
45.9	Chathill	54.00	50.5 ave	-			1/170 R, 1/150 F
65.6	Tweedmouth	75.05	56.3 ave	73.30	60.2 ave		L, 1/208 F, L
66.8	Berwick-on-Tweed	77.00		75.10		4¼ L	
83.2	Granthouse	103.45	36.8 ave	97.25	44.2 ave	6½ L	1/190 R, 1/200 R
87.9	Cockburnspath	109.15	51.2 ave	102.35	54.2 ave		1/200 R
95.2	Dunbar	115.55	65.8mph	109.10	66.4 ave	5¼ L	1/96 F, 1/200 F
106.8	Drem	-		120.25	61.9 ave	4½ L	undulating
117.8	Inveresk	139.45	57.0 ave	-			1/300 R, undulating
121.4	Portobello	143.15	61.7 ave	137.15	pws		1/300 F
124.4	Edinburgh Waverley	148.30	¼ E	143.15		5¼ L	

The superheated engines with reduced boiler pressure continued to give competent rather than sparkling performance on the heavier trains and 1244 is recorded as running the 44.1 miles from Darlington to York in 44 minutes 40 seconds with 365 tons and 1238 a load of 405 tons in 46 minutes. No detailed log is available for these but 1237 was timed on the northbound 9.13pm York–Newcastle, normally a 200 ton train hauled by an 'R'. On this occasion with 250 tons an 'R1' was provided. The second run beside this log was a special test of the new royal train vehicles plus dynamometer car.

At the Grouping there were still two at Neville Hill, 1243 and 1244, while 1242 had transferred to Starbeck. Because of less than satisfactory design of the ashpan, ash accumulation was excessive and coal consumption when worked hard was very high – they were nicknamed 'Miners' Friends'. Comparative power output calculations made by Professor Tuplin in his book *North Eastern Steam* over the Poppleton Junction–Eryholme section gave 'R1' 1244 a drawbar horsepower figure of 790 hauling a 365ton train averaging 67mph southbound, that is 30dhp persq ft of the grate area, a figure inferior to the performance of an 'R'.

Two 'R1s' were involved in train accidents in the second decade of the century – 1240 on the 4.05am mail from York to Newcastle in April 1912 was running over an hour late and the driver, attempting to make up time, caused the engine to derail because of excessive speed at Eaglescliffe on the route via Sunderland. Then in April 1918, 1244 derailed at Newby Wiske, the tender derailing first followed by most of the trailing coaches.

However, the arrival of the Gresley Pacifics at Gateshead displaced the Worsdell and Raven Atlantics ('V' and 'Z' – LNER 'C7' and 'C8') which in turn cascaded the D21s to semi-fast and other secondary duties. In 1925 the York D21s went to Neville Hill (1237, 1238 and 1240), while 1239, 1245 and 1246 went to Hull Botanic Gardens and 1241 joined 1242 at Starbeck. The Hull engines were usefully employed on the morning and afternoon heavy expresses for Liverpool as far as Sheffield. They were in turn displaced from these trains by the new Gresley K3 2-6-0s in 1931. The arrival of the D49s in the area made the D21s largely redundant from the early 1930s, though they were useful for some of the heavy summer holiday and excursion trains to the East Coast resorts.

As examples of their work in the 1930s, I table overleaf a couple of logs from the RPS archives on Leeds–York runs, the first a typical easy turn with light load and the second, a much better effort with a heavier load. Then there is an example of a Leeds–Hull turn, although the run was spoilt by continuous signal checks over the last ten miles. Apart from 1239's effort on the 5.30pm Leeds there was nothing requiring the potential power of the 'D21s', even in their dying days.

York–Darlington, c1909

		2137, 250 tons 9.13pm York–Newcastle		2137, 305 tons Empty Royal Train test		
Miles	Location	Times	Speed	Time	Speed	Gradients
0	York	00.00	T	00.00		
5.5	Beningbrough	08.10		08.53		L
9.6	Tollerton	12.10	62	-	56	L
11.1	Alne	13.35	64	14.53	63	L
13.1	Raskelf	15.35	60	16.40	66	1/666 R
16.1	Pilmoor	18.10	65	19.18	64	L
17.9	Sessay	20.00	64	21.07	65	L
22.1	Thirsk	23.40	68	25.33	sigs	L
26.5	Otterington	27.30	69½	30.35		1/629 R
29.9	Northallerton	30.40	64	34.14	55/60	1/671 R
33.7	Danby Wiske	34.00	68½	37.58	62	L
37.1	Cowton	37.00	68	-	60	1/650 R
39	Eryholme	38.35	71/64	43.34	56	1/490 R
41.5	Croft Spa	40.55	sigs	46.18		1/391 F
44.1	Darlington	44.20		¾ E 49.07	(46 net)	

Miles	Location	Leeds–York 1238, 6 chs, 145 tons 4.42pm Leeds 10.6.1933			1239, 10 chs, 250/265 tons 5.30pm Leeds 4.8.1934			Gradients
		Times	Speed		Times	Speed		
0	Leeds	00.00		T	00.00		T	
0.8	Marsh Lane	02.20			02.00			
2.7	Osmanthorpe	-			05.00	38		1/162 R
4.4	Cross Gates	08.00	38		07.25	42 ½		1/160 R
7.3	Garforth	13.20	sigs		11.05	52		1/160 R/ L
9.7	Micklefield	16.00	60		13.35	62		1/150 F
14.8	Church Fenton	20.55	62½		18.05	70		
16.5	Ulleskelf	22.55	54		19.45	62		L
17.9	Bolton Percy	24.15	58		20.55	68		L
21.8	Copmanthorpe	28.20	59		24.30	70		L
23.6	Chaloner's Whin	30.10	60		26.15 sig stop			
25.6	York	33.05	(32 net)	1 L	31.25 (29 net)		½ E	

Neville Hill's 1244 leaving York for Leeds on what looks like a Scarborough – Leeds service, 8 November 1923 just after repainting in LNER livery and the suffix 'D' put after the number. (H. Gordon Tidey/ MLS Collection)

		Leeds–Hull			
		1238, 6 chs, 187 tons			
		1.25pm Leeds–Hull			
		20.3.1937			
Miles	Location	Times	Speed		Gradients
0	Leeds	00.00		T	
0.8	Marsh Lane	02.25			
2.7	Osmanthorpe	05.30	37		1/162 R
4.4	Cross Gates	08.00	42½		1/160 R
7.3	Garforth	11.55	45/54		1/160 R, L
9.7	Micklefield	15.30	sigs		1/150 F
12.9	South Milford	20.10			1/136 F
16.5	Hambleton	24.15	57		L
18.3	Thorpe Gates	26.15	54		L
<u>20.7</u>	<u>Selby</u>	<u>29.35</u>	(27¼ net)	<u>2½ L</u>	
0		00.00		2½ L	
3	Hemingbrough	04.40			L
6	Wressle	08.00	54		L
11.7	Eastrington	13.30	65		L
14	Staddlethorpe	16.20	sigs 30*		L
16.5	Broomfleet	19.30	56		L
20.5	Brough	24.30	sigs 30*		L
		Sigs continuously for next 10 miles			
<u>31</u>	<u>Hull</u>	<u>43.10</u>	(35 net)	<u>6¾ L</u>	

Below left: **Starbeck's 1241** on a northbound 9-coach special working at Eryholme, 8 June 1930. (W. Rogerson/MLS Collection)

Below right: **York's 1239** leaving York and heading for Leeds with an express made up of very mixed rolling stock, c1925. Photographed from the Holgate platform. (F. Moore/MLS Collection)

Despite the apparent humdrum work of the ten members of the class in the late 1930s – or perhaps because of – they managed very respectable mileages which suggest they were robust engines with good reliability and availability, averaging 75,000 to 80,000 between heavy general repairs and over a million miles before withdrawal. The last few years of their career were spent exclusively at Neville Hill and Starbeck, the engines often exchanging base between the two locations. Their power was undoubtedly useful during the war years and one, 1242, was attacked when working a Hull–Scarborough train and bombed, the bomb falling clean through the middle of the tender. They were allocated numbers 2217–2223 in the proposed LNER 1946 renumbering scheme, but they were all withdrawn before the renumbering could take place. After withdrawal in July 1943, 1241 was despatched to Fishguard for RAF target practice and its shell was returned in January 1944 to Darlington for scrapping.

The NER 'D', 'F' (LNER D22)

We now take a step backwards in time to the days of Thomas Worsdell. In the chapter about McDonnell's class '38', I described the problems he encountered and his resignation in 1884. The gap he left was covered for a year by the General Manager, Henry Tennant, with a committee from the locomotive department. During that year, twenty 2-4-0s were designed and put into traffic and performed satisfactorily on express work. Thomas Worsdell was recruited from the Great Eastern and his first priority was to sort out the capacity problems of Gateshead Works which had developed a large backlog of repairs. His initial locomotive designs were tank and freight engines, but by the late 1880s passenger loads were stretching the Fletcher and Tennant 2-4-0s and the McDonnell 4-4-0s had shown themselves incapable of developing the power now required for main line express work.

In 1886, Worsdell produced a 2-4-0, numbered 1324, which was identified as class 'D', a development of the Tennant 2-4-0 with Worsdell-von Borries compounding and a larger boiler and a deeper sloping grate. Boiler pressure was 170lbs psi, later raised to 175. Other dimensions of the 2-4-0 were:

T.W. Worsdell's class 'D' compound 2-4-0 1324 of 1886 at Newcastle for the Queen's Jubilee Exhibition, 1887. It was rebuilt as a simple expansion 4-4-0 in 1896 as a class 'F' and withdrawn in 1930 as an LNER D22. (F. Moore/MLS Collection)

Cylinders (2):	High pressure 18 x 14in
	Low pressure 26 x 24in
Coupled wheel diameter:	6ft 8in
Leading wheel diameter:	4ft 7¾in

Joy's valve gear with slide valves

Heating surface:	1,323sq ft
Grate area:	17.3sq ft
Axleload:	17 tons 19 cwt
Weight (Engine):	43 tons 7 cwt
(Tender):	35 tons 8 cwt
(Total):	78 tons 15 cwt

In 1887, ten compound 4-4-0s followed of similar design but with a leading bogie, as already there had been some concern about the potential unsteadiness at speed with 1324, the weight distribution caused by the difference in cylinder size. (As 1324 had been on static exhibition at Newcastle in connection with Queen Victoria's Golden Jubilee there was no actual evidence of this.) They were allotted blank numbers in the NER locomotive fleet, 1, 18, 42, 115, 117, 355, 514, 663, 684 and 779. No.1 was renumbered 356 in 1914 when electric locomotive No.1 for the NER was constructed. The bogie wheels were 3ft 7¼in and the weight increased by just over three tons. They were designated class 'F'. One of them, 117, demonstrated its speed capabilities by running from Newcastle to Edinburgh in 126 minutes on the final day of the August 1888 'race' to Aberdeen with the rival LNWR company.

At the same time, Gateshead constructed ten similar 'simple expansion' 4-4-0s for comparative purposes, though they initially only had a 140lbs psi boiler compared with 175 of the compounds. Axleload was reduced to 17 tons exactly. They otherwise had similar dimensions and weighed half a ton less. They were numbered 85, 96, 154, 194, 230, 673, 777, 803, 808 and 1137 and were designated class 'F1'.

In 1888, a further class 'D' 2-4-0 emerged new from Gateshead Works. It would seem strange to build another 2-4-0 at this stage, but it differed in having inside admission piston valves invented by W.M. Smith, Gateshead's Chief Draughtsman. The novelty of the design may have delayed its construction so that it fell behind the delivery of the 4-4-0s. It was numbered 340 and also classified as 'D', despite its differences to 1324.

Experience with the two classes of 4-4-0s gave the compounds an advantage of approximately 15 per cent lower coal consumption, and fifteen more were constructed in 1890 and 1891 as class 'F' and numbered in a straight series, 1532–1546. In the interim period between the building of the two series of 'F' 4-4-0s, T.W. Worsdell had built two small classes of 4-2-2s for express passenger work, but it was the 4-4-0 that was seen as the way forward.

Wilson Worsdell took over from his brother in 1890 and designed and constructed twenty larger 4-4-0 simple expansion engines (the 'M', LNER D17/1) and one '3CC' (LNER D19), described earlier. In 1893 there was an investigation into the performance of the simple and compound locomotives and despite the alleged superior coal consumption of the compounds, presumably because of the more costly maintenance of the complex compound machines, it was decided that the 'F' compounds should be rebuilt as simple expansion engines. Wilson Worsdell decided to rebuild the two class 'D' 2-4-0s also in the same way as 4-4-0s and the entire fleet of class 'D', 'F' and 'F1s' were brought into one class of 4-4-0s between 1896 and 1905. The 'F1s' were rebuilt with Stephenson motion and piston valves so that all engines were now identical and from 1914, were designated 'F'. Between 1913 and 1920 they were all equipped with superheaters and passed into the LNER locomotive stock as class D22. Their dimensions at the Grouping of this class of thirty-seven locomotives were:

Cylinders (2-inside):	18 x 24in
Coupled wheel diameter:	6ft 8in
Bogie wheel diameter:	3ft 7¼in

'F' compound 779, seen at Edinburgh Waverley, c1900. It was rebuilt as a simple expansion engine with piston valves in 1902. (Real Photographs/ MLS Collection)

Stephenson motion with 7½in piston valves
Boiler pressure: 160lbs psi
Heating surface: 1,092sq ft, incl 185sq ft of superheating
Grate area: 17.3sq ft
Axleload: 17 tons 6 cwt
Weight (Engine): 48 tons 8 cwt
(Tender): 37 tons 10 cwt
(Total): 85 tons 18 cwt
Water capacity: 3,038 gallons
Coal capacity: 5 tons
Tractive effort: 13,210lbs

The livery of the 'Fs' was identical to the other NER 4-4-0s – NER green, followed by LNER lined apple green after the Grouping and black with red lining from 1928. No.96 was withdrawn after a head-on collision at Hull Paragon station in 1927, and after the introduction of the D49s causing a cascading of NER 4-4-0s to inferior work, the D22s found themselves redundant and all were withdrawn between 1929 and 1935. The last survivors were 1546 withdrawn in January, 777 withdrawn in May, and 1537 withdrawn in November 1935.

Operation

The twenty compounds and simple expansion 'Fs' and 'F1s' were intended for the main line and matched the Tennant '1463' 2-4-0s that were the current mainline engines with the earlier Fletcher 2-4-0s – indeed, the compounds were free runners once underway and were key performers in the 1888 rivalry with the West Coast LNWR and Caledonian Railways in their effort to reach Aberdeen first. However, they were not as reliable, and their cost of maintenance

Above: Former 'F' compounds 42, rebuilt in 1904 and 356 (the former No.1) rebuilt in 1903, on Scarborough shed together in LNER lined green livery, c1926. (MLS Collection)

Below: Former 'F' compound 663, rebuilt in 1900, in LNER black livery, at Hull Botanic Gardens shed, c1930. It was withdrawn in October 1932. (Real Photographs/MLS Collection)

1324, the former 'D' compound 2-4-0 of 1886, rebuilt as a 4-4-0 with Stephenson gear and piston valves in 1896 on a ten coach semi-fast express near Benton Quarry, 29 August 1919.
(K. Nunn/LCGB/MLS Collections)

offset any advantage of lower coal consumption. They had to make way for the Wilson Worsdell 'M' and 'M1' classes from 1892 and were displaced entirely once the 'Q' engines came on stream in 1896.

The 'Fs' were involved in a number of train accidents. In 1889, 663 was a runaway on Seaton Bank when working the 2pm Liverpool–Newcastle express. There was a 1in 39-44-60 descent and the driver, attempting to regain some of the train's 16 minute lateness, passed the top of the bank at 40mph and the application of the Westinghouse brake was insufficient to prevent derailment at the foot of the bank on the 15mph restricted curve at Ryhope. In 1903, 777, after passing a signal at danger, was derailed at York. This engine, based later at Hull was reputed to be the 'black sheep' of the class. 'F' 85 was involved in an accident in November 1906 at Ulleskelf when working a Scarborough–Leeds train. As sometimes happened, it was racing a L&Y train on the four track section, but unusually, because of freight train congestion ahead, the signalman diverted the Leeds train and it smashed into the back of a goods at an estimated 50mph, killing the crew.

Their rebuilding as the standard piston valve simple expansion 'F' class between 1896 and 1905 made them competent power for secondary services working from Newcastle to Carlisle, Leeds–Scarborough and Leeds–Hull before the Grouping.

The allocation at the Grouping was:

Alnmouth:	18
Carlisle:	154, 355, 1324
Gateshead:	1546
Waskerley:	663
Starbeck:	1533, 1545
Scarborough:	356, 779, 1541, 1542
Selby:	673, 1532, 1535, 1537, 1544
Springhead:	808, 1538

The remainder were concentrated at Bridlington (6) and Hull Botanic Gardens (11). Noted transfers in

'F' compound 1540 at York with a southbound stopping passenger train, c1900. It was rebuilt as a simple expansion engine with piston valves in July 1905. An intriguing early motor vehicle is on the flat wagon behind the engine – an early motorail service!
(MLS Collection)

the 1920s before their withdrawals began in earnest at the end of the decade included:

- 18 to Tweedmouth, then to Gateshead in 1928
- 808 to Gateshead in 1925 and Stockton in 1927
- 663 and 1546 to Hull Botanic Gardens in 1929
- 777 to Waskerley in 1930
- 1542 to Stockton in 1930
- 1541 to Starbeck in 1925
- 1538 to Scarborough in 1924
- 115 from Hull to Starbeck in 1931

There was a shortage of power on the former Great Eastern lines in 1926 and 777, 1540 and 1544 together with three D17/2s were transferred there until arrears of maintenance at Stratford were recovered. 777 and 1544 were allocated to Ipswich and 1540 to Cambridge. All three had returned to the North East by April 1927.

The last D22 in service was 1537 at Selby, withdrawn in November 1935.

The NER 'G' (LNER D23)

Thomas Worsdell matched his class 'F' express passenger engines with a smaller wheeled engine – with 6ft 1¼in coupled wheels instead of 6ft 8in - for secondary passenger and mixed traffic use, NER class 'G1' ('G' was retained for a possible compound version). The twenty locomotives, numbered 23, 214, 217, 222, 223, 258, 274, 328, 337, 372, 472, 521, 557, 675-679, 1107 and 1120, were built at Darlington in 1887 and 1888 and started life as 2-4-0s however, and it was only under his successor, Wilson Worsdell, that they were rebuilt as 4-4-0s.

Above: **T.W. Worsdell** 'G1' 2-4-0 676 built in 1888 and rebuilt as a 4-4-0 in 1904, seen at Ordsall Lane depot, c1900. It was withdrawn as an LNER D23 in 1930. (MLS Collection)

Below: **223**, G1 2-4-0 rebuilt as a 'G' 4-4-0 in June 1901, seen here c1905: It was superheated in 1914 and withdrawn in 1933. (Locomotive Publishing Co./MLS collection)

217, rebuilt from the 1888 2-4-0 in 1901, superheated in 1914, at Middleton-in-Teesdale, c1928. (N. Fields/MLS Collection)

The dimensions of the locomotives as built were as follows:

Cylinders (2-inside):	17 x 24in
Coupled wheel diameter:	6ft 1¼in
Leading wheel diameter:	3ft 1¼in
Joy's motion with slide valves	
Boiler pressure:	160lbs psi
Heating surface:	1,092sq ft
Grate area:	15.6sq ft
Axleload:	16 tons 2 cwt
Weight (Engine):	40 tons 4 cwt
Water capacity:	2,651 gallons
Coal capacity:	4 tons

Between 1900 and 1904, Wilson Worsdell rebuilt all twenty engines as 4-4-0s with Stephenson motion

472, rebuilt from 2-4-0 in 1903, at Darlington, 1930. Note number now on tender. (W. Leslie Good/MLS Collection)

and piston valves. The altered dimensions were:

Cylinders (2-inside):	18 x 24in
Bogie wheel diameter:	3ft ¼in
Heating surface:	1,082sq ft
Axleload:	15 tons 13 cwt
Weight (Engine):	44 tons 7 cwt
(Tender):	34 tons 16 cwt
(Total):	79 tons 3 cwt
Tractive effort:	14,437lbs
Water capacity:	3,038 gallons
Coal capacity:	5 tons

As no compound version was built, the class was redesignated as 'G' in 1914. All were further rebuilt with superheated boilers between 1913 and 1916, with a heating surface of 1,001sq ft including 175sq ft of superheating. A longer smokebox was then used to encompass the superheater. They were reliable but unspectacular engines, nicknamed 'Waterburys' after a common cheap but reliable watch of the period.

At the Grouping, the twenty engines were reclassified as D23. The NER had painted them in the passenger green livery but since they were not seen as express locomotives by the LNER locomotive management, they were painted black with red lining from the start. The first withdrawal was of No.23 in August 1929 and the last survivor was 1120, withdrawn in May 1935, having outlived its sisters by nearly two years.

Operation

As 2-4-0s these twenty locomotives worked semi-fast and stopping passenger services radiating from Hull, Leeds, Harrogate and Scarborough and after their rebuilding in the early 1900s continued on the same work. In 1907 their distribution was:

Hull Botanic Gardens:	16
Starbeck:	3
York:	1

The Botanic Garden engines worked nearly all the passenger trains out of Hull except the fastest expresses and the Starbeck three worked to Leeds and Bradford, West Hartlepool and Saltburn. The solitary York engine was used for the Malton and Whitby line. They were reliable engines and availability on this relatively light work was high, so they ran good annual mileages. The type of work on which they were employed seldom called for much power or speed, rarely reaching 60mph, for instance, 223 on a 4-coach Hull–Bridlington train ran the 31 mile route in 37 minutes 50 seconds with a minimum of 27mph on Flamborough Bank with 58mph on the descent. 521 was recorded at 71mph on a Bridlington–York train recorded in a *Railway Magazine* article by Cecil J. Allen, the highest known speed of a 'G'. The known timings were:

Bridlington–York

521 'G', 5 chs, 90 tons

6.45pm Bridlington–York

Miles	Location	Times	Speed		Gradient
0	Bridlington	00.00		T	
	Nafferton	-	sigs		L
11.5	Driffield	16.45	slack*		
18.7	Middleton-on-the-Wolds	-	45		1/124 R
21.5	Enthorpe	33.30	27		1/97 R, 1/156 R
25.2	Market Weighton	38.00	60/45*		1/95 F
	Pocklington	-	30 poor steaming		
35.7	Fangfoss	51.35	50		
45	Earswick	60.05	68/71½		
47.5	York	64.00	(63 net)		4 L

217 on a local passenger train at Middleton–in-Teesdale, c1928. (MLS Collection)

At the Grouping they were still undertaking the same type of work although the Hull fleet had reduced to eleven, with Starbeck and Scarborough having three each and the remainder were at Bradford, Malton and Darlington. However, under LNER management and an influx of new locomotives causing a cascade of consequences, the 1925 allocation had changed and by 1928 had changed again to:

	1925	1928
Hull Botanic Gardens:	6	2 (677, 1120)
Leeds Neville Hill:	1	-
Starbeck:	2	-
Manningham:	1	1 (328)
Darlington:	3	8
Barnard Castle:	3	3
Middleton-in-Teesdale:	1	1
Tweedmouth:	1	1 (223)
Duns (ex NBR):	2 (223, 258)	1 (258)
Malton:	-	1
Kirkby Stephen:	-	1 (678)
Waskerley:	-	1 (23)

Withdrawals began with 23 in August 1929 and by the end of 1931 only five remained and were concentrated on the Stainmore line – 223, 337, 677, 678, 1120. With the withdrawal of 328 from Manningham at the end of 1931, and a D22 115 in 1933, 1120 replaced it working Bradford–Harrogate–York services until its withdrawal as the last working D23 in May 1935.

The Hull & Barnsley Rly Class 'J' (LNER D24)

The Hull and Barnsley Railway (H&BR) was opened in 1885, but it was not until 1905 that the company was able to run passenger services through from Hull to Sheffield by way of obtaining running powers from Cudworth to Sheffield Midland station. In 1910. the company's life expired 2-4-0s used on passenger services were replaced by five new 4-4-0s built by Kitson & Co. They were given five numbers of the previous 2-4-0s, 33, 35, 38, 41 and 42. Their main dimensions were:

Cylinders (2-inside):	18½ x 26in
Coupled wheel diameter:	6ft 6in
Bogie wheel diameter:	3ft 9in
Stephenson motion with slide valves	
Boiler pressure:	175lbs psi (later 170lbs psi)
Heating surface:	1,309sq ft (later 1,398sq ft)
Grate area:	19.6sq ft (later 19.4sq ft)
Axleload:	18 tons
Weight (Engine):	55 tons 2 cwt
(Tender):	40 tons 10 cwt
(Total):	95 tons 12 cwt
Water capacity:	3,500 gallons
Coal capacity:	5 tons 5 cwt
Tractive effort:	16,970lbs (later 16, 485lbs)

The H&BR was taken over by the North Eastern Railway in 1922 and

Hull & Barnsley Railway 4-4-0 No, 38 shortly after construction, c1912. (F. Moore/MLS Collection)

Hull & Barnsley Railway No. 33 at Sheffield Midland, c1919. (W. Leslie Good/MLS Collection)

3000 added to the former number, although only 3038 was turned out in NER green livery. The other engines retained their green H&BR livery with their new numbers on the cab until repainted in LNER livery, 3041 the first in May 1923. However, in 1924 the LNER renumbered them 2425–2429 in the batch of numbers allocated to NER 4-4-0s to avoid their duplication with GNR locomotives in the LNER 3000 series. Like other LNER 4-4-0s, they lost their lined apple green livery in 1928 to black with red lining.

The locomotives were due for reboilering in 1929-30 and, surprisingly for a small class of non-standard engines, they got domed saturated steam boilers with slightly changed dimensions – as indicated in the dimensions listed above. However, this failed to

D24 2429 (ex H&BR 42) seen in its last year of life, with the domed boiler fitted in 1929. (MLS Collection)

prolong their lives to any extent and after the Sheffield service was cut back from Cudworth to South Howden in 1932, they spent most of their time in store, being withdrawn in 1933 and 1934, the last survivor being 2429 withdrawn in September 1934.

Operation
They were built for the Hull (Cannon Street)–Sheffield Midland service which consisted of three services each way on a weekday. They were occasionally used for football excursions and the twice-yearly employee excursions to Blackpool, Morecambe or Llandudno. The Sheffield services were cut back to Cudworth in 1917 and in 1924 the trains were run into Hull Paragon station instead of Cannon Street.

In 1932, the service was further reduced with the former H&BR service terminated at South Howden when it became a nonsense to use these large 4-4-0s on such a truncated service. They were then used on other secondary services to Doncaster or York or Bridlington, and 2425 and 2427 spent a short time in 1931 on loan to Selby. However, they were clearly redundant by 1933 and spent time in store, 2425 being withdrawn in August 1933 despite its reboilering in 1929. 2426 and 2428 were withdrawn in December, 2427 in January 1934 and 2429 in September. In fact, their boilers were retained as stationary boilers, two at Springhead, two at Stratford and that from 2428 reused on a J28 0-6-0.

A rare shot of a D24 in action - 2429 on the 2.55pm Sheffield–Cudworth, 1934. (MLS Collection)

Chapter 3
THE NORTH BRITISH 4-4-0s

The chronological order of the construction of the North British 4-4-0s was not followed by the LNER in classifying the different groups of engines, and whilst I deal with the classes in the LNER order for easy reference, for clarity I list below the classes in the order in which they were designed, built and rebuilt.

The NBR 'Letter' classification does not seem to have been used in common parlance and either the number of the first member of the class, description or appropriate class name used.

LNER class	NBR class	Wheel Diameter	Built	Designer	Rebuilt
D27	M '476' Abbotsford'	6' 6"	1877 – 1879	Drummond	1902 Holmes
D28	M '476 Abbotsford'	6' 6"	1877 – 1879	Drummond	1904 Reid
D31	M '574', '633' & '729'	6' 6"	1884 – 1899	Holmes	
D25	N '592'	7' 0"	1886 – 1888	Holmes	
D35	N 'West Highland Bogies'	5' 7"	1894 – 1896	Holmes	
D36	L ' Rebuilt West Highland Bogie'	5' 7"	1894	Holmes	1919 Reid
D26	K '317'	6' 6"	1903	Holmes	
D32	K 'Intermediate'	6' 0"	1906 – 1907	Reid	
D29	J 'Scott'	6' 6"	1909 – 1911	Reid	
D33	K 'Intermediate'	6' 0"	1909 – 1910	Reid	
D30	J 'Superheated Scott'	6' 6"	1912 – 1920	Reid	
D34	K 'Glen'	6' 0"	1913 – 1920	Reid	

The NBR Class 'N' (LNER D25)

Matthew Holmes followed Dugald Drummond at Cowlairs in 1882, when the company was relying on his predecessor's dozen 4-4-0s for main line express train working, supplemented by Wheatley's 2-4-0s and 4-4-0s, none of the latter surviving to the Grouping. Holmes designed and constructed twelve 4-4-0s with 7ft coupled wheels which were built at Cowlairs between 1886 and 1888, numbered 592-603. The intention had been to use them on through working to Aberdeen with the re-opening of the Tay Bridge, which took place in June 1887. Their dimensions were:

Cylinders (2-inside):	18 x 26in
Coupled wheel diameter:	7ft 0in
Bogie wheel diameter:	3ft 6in
Boiler pressure:	150lbs psi

Stephenson motion with slide valves

Heating surface:	1,106.18sq ft
Grate area:	21sq ft (later 20.3sq ft)
Axleload:	17 tons 2 cwt

592 was displayed at the Edinburgh International Exhibition in the summer of 1886 and is seen here in front of the Scott memorial, 1886. (Real Photographs/ MLS Collection)

602 was decorated with the Prince of Wales's feathers for the opening of the Forth Bridge and the running of the royal train with Prince Edward, 4 June 1890. (MLS Collection)

The North British 4-4-0s • 67

9595 after removal of smokebox wingplates and equipping with standard NB cab, at Haymarket shed, c1928. 9595 was withdrawn in March 1932. (MLS Collection)

Weight (Engine):	45 tons 5 cwt (later 46 tons 4 cwt)
(Tender):	32 tons
(Total):	77 tons 5 cwt (later 78 tons 4 cwt)
Water capacity:	2,500 gallons
Coal capacity:	4 tons
Tractive effort:	12,786lbs

593 was fitted with an extended smokebox in 1901 which it retained until it, along with other members of the class, was reboilered in 1911. All had smokebox wingplates even after reboilering, although these were removed in the early years after the Grouping. Their Holmes rounded cab outline was replaced by Reid with side-window cabs that became the North British standard.

As with other North British engines, 9000 was added to their numbers in 1924/5 making them LNER 9592–9603, though 593, 594, 597, and 602 were withdrawn in 1926 before receiving their new numbers. Between 1923 and their renumbering, the letter 'B' was added after the number. The entire class was repainted in the LNER apple green livery and those still operating in 1928 were then repainted black with red lining. When built, the locomotives were fitted with the Westinghouse air brake system, though by 1908, five

Below left: At the turn of the century, the NB regular royal engine was 594, seen here decorated with the crown painted on the splasher, ready to go off shed with a train for Aberdeen, c1900. (Photomatic/MLS Collection)

Below right: 603 seen on shed, c1905. (Photomatic/MLS Collection)

595 hurries a two Midland Railway coach portion of an express from London between Perth and Aberdeen, c1905. (G.M. Shoults/ MLS Collection)

were also vacuum fitted. All were dual braked by 1920.

As stated earlier, the first withdrawals took place in 1926, a couple, 9600 and 9603, were withdrawn in 1928, two more, 9598 and 9599 in 1930, 9592 and 9595 in 1932 and the last survivor, 9596, in July 1933.

Operation

Initially they were used on Edinburgh–Aberdeen expresses as far as Burntisland until the Forth Bridge was opened in 1890 after which they could work through to Aberdeen. They ran the expresses between Edinburgh and Glasgow. They also worked south of Edinburgh as far as the border at Berwick-on-Tweed.

By the end of the first decade of the twentieth century, they were relegated to stopping train services or occasional piloting. At the Grouping their allocation was:

Glasgow Eastfield:	594, 600, 602, 603
Haymarket:	595 – 597
Stirling:	592, 601
Dunfermline:	593
Dundee:	598
Hawick::	599

603 had a spell on the West Highland line at Fort William and 9601 was later sub-shedded at Dunbar for local services and banking work on the climb to Grantshouse. 9592 moved to Eastfield and was regularly used as a pilot engine at Cowlairs. Eastfield's last route work for the D25s was from Glasgow Queen Street to Kinross via Larbert and

An unknown D25 is placed inside D11/2 6394 leaving Aberdeen with the 2.30pm to St Pancras, 16 September 1927. (K. Nunn/LCGB/ MLS Collections)

Alloa. The Haymarket engines, as well as local stopping services, were frequently used for piloting main line services to Glasgow or Dundee. 9603 was transferred to Hawick and also spent time banking trains to Whitrope summit as well as other local duties. The suitability of engines with 7ft coupled wheels is questionable but their work at Cockburnspath, Cowlairs, Inverkeithing and Riccarton seems to have been satisfactory as they remained doing this in their latter days. The last survivor, 9596, had been in store at Dunfermline since its banking work at Inverkeithing finished in September 1931, so the last operating D25s were 9592 and 9595 at Eastfield and Haymarket respectively.

The NB 'K' (LNER D26)

Matthew Holmes designed and built a dozen 4-4-0s with 6ft 6in wheel diameter under the NBR directors' 1902 authority from May 1903, although unfortunately only four of these were in traffic before Holmes was taken ill and died in July. They were called the '317' class, numbered 317–328, and had larger cylinders and piston valves instead of slide valves of earlier NB passenger engines. Although built with the typical Holmes rounded cab, they were altered to give a side-window cab in the form that Holmes' assistant, W.P. Reid, provided for his own engines. They were constructed at Cowlairs and their dimensions were:

Cylinders (2-inside)	19 x 26in
Coupled wheel diameter:	6ft 6in
Bogie wheel diameter:	3ft 6in
Boiler pressure:	200lbs psi (later reduced to 190lbs)

Stephenson motion with piston valves

Heating surface:	1,577sq ft
Grate area:	22.5sq ft
Axleload:	18 tons 6 cwt
Weight (Engine):	52 tons
(Tender):	40 tons
(Total):	92 tons
Water capacity:	3,525 gallons
Coal capacity:	7 tons
Tractive effort:	19,434lbs

Despite Reid's introduction of more powerful Atlantics in 1906, they held sway on the NB's main line services, and were seen to be both successful and popular although only these twelve were built. Differing from normal practice, they were not rebuilt when due for reboilering as their frames were in poor condition and three – 317, 320 and 328 – were withdrawn as early as July–November 1922. The remainder were renumbered with the addition of 9000, but only one, 320, was painted in LNER lined green livery with the NB number and 'B' suffix, the rest when renumbered in 1924. 323 was withdrawn in 1923, 317 in 1924 and all the rest were withdrawn, 9325 being the only one to survive to 1926 (July).

Five were fitted with Midland style steam heating apparatus for working with ex-Midland stock over the Waverley route and another five with tablet catchers for the Dundee–Aberdeen main line. All were built with smokebox wingplates but only the three withdrawn in 1922 and 317 withdrawn in 1924 retained them to the end. 317-322 had the Westinghouse brake system, the others were dual braked.

Holmes Class 'N' 319 as built with Holmes cab, before Reid equipped the class with side-window cabs, c1910. 319 was one of the first of the class to be withdrawn in 1922 before the Grouping. (MLS Collection)

320, now equipped with a Reid side-window cab, leaving Aberdour, August 1911. (G.M. Shoults/MLS Collection)

Operation

The first five were allocated to Aberdeen, replacing the class 'N' 7ft 4-4-0s of 1886. Their primary duties were the Edinburgh expresses with 323 and 324 performing the same function from St Margaret's depot. 325–328 were also based at St Margaret's but were for the Waverley route to Carlisle. Then to balance the working, 327 and 328 were based at Carlisle for early trains originating at that end of the route. In 1906, the 'Ks' were themselves replaced by the first Reid Atlantics, being transferred to Dundee and St Margaret's, the latter depot's duties now including expresses to both Glasgow and Perth. By the First World War they had no regular diagrams but were in demand for special traffic or replacing booked engines at short notice.

Holmes Class 'N' 319 working a stopping train from Carlisle to Silloth, c1910. (MLS Collection)

Above left: 322 with a northbound express in the Aberdour area, c1911. (G.M. Shoults/MLS Collection)

Above right: 327 tops the rise near Dalgetty Box with the 7.40am Edinburgh Waverley–Aberdeen, 11 August 1911. (G.M. Shoults/MLS Collection)

Towards the end of their lives, they were frequently used for piloting heavy trains as well as use on stopping services – in the immediate post-war situation train loads on both the Edinburgh–Aberdeen and Edinburgh–Carlisle routes required frequent double-heading.

After the Grouping, 320B was based at Hawick for a few months before its withdrawal while the other six went to Bathgate working Edinburgh–Glasgow via Airdrie trains.

The NB Class 'M' (LNER D27 & D28)

We go back a generation to the era of Dugald Drummond at Cowlairs. Eight 6ft 6in wheeled 4-4-0s to that engineer's design delivered by Neilson & Co. in 1877 and 1878 were supplemented by a further four built at Cowlairs in 1879. The Neilson engines were numbered 476-479 and 486-489 and the Cowlairs locomotives, 490-493. They were named:

476 *Carlisle*
477 *Edinburgh*
478 *Melrose*
479 *Abbotsford*
486 *Aberdeen*
487 *Montrose*
488 *Galashiels*
489 *Hawick*
490 *St.Boswell's*
491 *Dalhousie*
492 *Newcastleton*
493 *Netherby*

With the transfer of 486 and 487 to Carlisle after the collapse of the Tay Bridge in 1879, they were renamed *Eskbank* and *Waverley*, but all the names were removed after Holmes replaced Drummond in 1884.

They were a significant design for Drummond himself, large by contemporary standards and precursors of a generation of 4-4-0s, not only for the North British Railway but for his influence on later designs for other Scottish railways and the London & South Western too. They became known as the 'Abbotsfords' as the official photograph of the Neilson company publicising the design depicted 479. Their dimensions were:

Cylinders (2-inside):	18 x 26in
Coupled wheel diameter:	6ft 6in
Bogie wheel diameter:	3ft 6in
Boiler pressure:	150lbs psi
Stephenson motion with slide valves	
Heating surface:	1,099.3sq ft
Grate area:	21sq ft
Axleload:	15 tons 10 cwt
Weight (Engine):	44 tons 5 cwt
(Tender):	32 tons
(Total):	76 tons 5 cwt
Water capacity:	2,500 gallons
Coal capacity:	6 tons
Tractive effort:	13,770lbs

After twenty-five years of hard work on the NB's main lines, six were rebuilt with larger boilers by Holmes in 1902 and the other six similarly by Reid in 1904. The cylinder diameter was slightly enlarged to 18¼in and

Drummond 'M' class 491, formerly named *Dalhousie*, built at Cowlairs in 1878, before rebuilding by Reid in 1904, with the original Drummond cab, c1900. (Rail Archive Stephenson/John Scott-Morgan Collections)

the boiler pressure raised to 175lbs psi, the heating surface to 1,350sq ft, though the grate area was reduced to 20sq ft. The engine weight increased by a couple of tons. The only difference between the Holmes rebuilds and those of Reid was the cab design. The Holmes engines received his curved pattern while the Reid rebuilds got the single side-window cab that featured in the '317' class introduced the previous year.

In 1919, the twelve locomotives were placed on the duplicate list and were renumbered:

Holmes rebuilds	Reid rebuilds
476 to 1321	477 to 1322
478 to 1323	486 to 1360
479 to 1324	487 to 1361
488 to 1362	491 to 1387
489 to 1363	492 to 1388
490 to 1371	493 to 1389

At the Grouping, the LNER classified the Holmes rebuilds as D27 and the Reid's as D28. The smokebox wingplates were removed, four in 1921 and the rest after the Grouping. The following engines were withdrawn during the two years preceding the Grouping – 1360 (ex 486), 1362 (ex 488), 1363 (ex 489), 1371 (ex 490), 1389 (ex 493). 1324 (479) was withdrawn in 1923, and the survivors were allocated new LNER numbers as follows:

NB	NB duplicate	1st LNER	Final LNER
476	1321	10321	9992
477	1322	10322	9993
478	1323	10323	9319
479	1324	10324	-
487	1361	10361	9994
491	1387	10387	9995
492	1388	10388	9996

Only 10361 and 10387 carried these numbers and none of the engines survived long enough to carry the second LNER allotted number as all apart from these two had been withdrawn by 1924 and the last two were retired in September 1926.

Operation

The entire class (apart from two of the 1878 Cowlairs built engines) was initially allocated to St Margaret's and Carlisle to run the Midland expresses over the Waverley route. This difficult route, 98 miles with the climbs in both directions to Falahiill and Whitrope, was scheduled in 140 minutes for a non-stop train and 155 minutes for a train with three stops, timings hardly improved on in later timetables. The 'Abbotsfords' performed well and coal consumption was very economical, said to be just 28lbs per mile in 1878 when the average load was about 120 tons.

The other two locomotives, 486 and 487, were allocated to Aberdeen and worked the Edinburgh services to Burntisland (before the building of the Forth Bridge) until 1879 and the collapse of the Tay Bridge. They then joined their sisters on the Waverley route. The St Margaret's engines also worked to Glasgow, Perth and Dundee. One for many years was used to double-head trains from Edinburgh as far as Falahill summit and then return light engine to Edinburgh to assist the next southbound service.

After rebuilding in 1902/4 they had been displaced by the Holmes '317' class and a couple of years later by the Reid Atlantics, and six – 478, 479, 487, 491-493 – were allocated to Eastfield for work on the West Highland line, some being sub-shedded at Fort William. However, they were in turn replaced by Reid's NBR 'K' class of 6ft 4-4-0s (LNER D32) around 1910 and were employed chiefly on stopping trains to Hawick on the Waverley, Glasgow, Perth and Dundee. In July 1916 487 was commandeered to work a one coach special to Carlisle and completed the non-stop run in 104 minutes, a time that has never been bettered – at least not in steam days.

I have found one surviving log from a run with the first example of the class on a Glasgow–Edinburgh/Kings Cross train in the latter part of the first decade of the twentieth century – the load was a heavy one for an 1878 built 4-4-0, even one rebuilt by Reid in 1902 :

Glasgow Queen St–Haymarket, c1906-1910

476 (formerly *Carlisle*)

8 chs, 250 tons

8.45am Glasgow–King's Cross

Miles	Location	Times	Speed	Gradients
0	Glasgow Queen ST	00.00		1/41 R
1.5	Cowlairs	06.20	(banked)	
6.3	Lenzie	12.50	46	L
11.5	Croy	19.15	50	1/900 R
21.8	Falkirk High	30.15	57	L
25	Polmont	34.20	47	L
29.7	Linlithgow	40.05	50	1/882 F
39.1	Ratho	50.55	53	1/960 F
43.8	Saughton Junction	55.55	57	1/960 F
46.1	Haymarket	58.45	1¼ E	

Drummond 'M' 476 *Carlisle* before rebuilding by Holmes in 1902 on a Waverley route express in the Edinburgh area, c1900. (G.M. Shoults/MLS Collection)

Drummond 'M'
479, formerly named *Abbotsford*, rebuilt by Holmes in 1902 with a Holmes style cab, and transferred from Eastfield to St Margaret's in 1910, seen on a stopping train in the Edinburgh area, 1911. 479 was LNER D27 and was withdrawn in December 1923.
(G.M. Shoults/MLS Collection)

The allocation of the surviving engines at the Grouping was:

Eastfield: 1321, 1322, 1388
Haymarket: 1323, 1324, 1361, 1387

The last survivors were 10361 working local services around Glasgow, presumably reallocated to Eastfield, and 10387 at Haymarket working Edinburgh–Dundee stopping trains via Crail.

The NBR Class 'J' (LNER D29)

Reid had introduced a 4-4-0 'Intermediate' with 6ft coupled wheels in 1906, which, although built for goods work, performed well on passenger services also. As passenger train loads were increasing, further 4-4-0s were needed and for express work Reid designed what became known as the 'Scott' class, NBR class 'J', with 6ft 6in coupled wheels. A large tender was provided to enable non-stop working of the Midland expresses from Carlisle to Edinburgh. Six, numbered 895–900, were built by the North British Locomotive Company at their Hyde Park Works and a further ten were built at the NB's Cowlairs Works in 1911, numbered 243–245, 338–340, and 359–362. Their key dimensions were:

Cylinders
  (2 – inside):   19 x 26in
Coupled wheel
  diameter: 6ft 6in
Bogie wheel
  diameter: 3ft 6in
Boiler pressure: 190lbs psi

Stephenson motion with piston valves
Heating surface: 1,618.12sq ft
Grate area: 21.13sq ft
Axleload: 18 tons 8 cwt

Weight (Engine):   54 tons 16 cwt
       (Tender):   46 tons
       (Total):   100 tons 16 cwt
Water capacity: 4,235 gallons
Coal capacity: 7 tons

During the era, naming of locomotives had not been pursued – in fact names were removed from the Drummond 'Abbotsfords' – but Reid reintroduced naming on his 1906 Atlantics and these 4-4-0s were named after characters from Sir Walter Scott's novels:

895	*Rob Roy*
896	*Dandie Dinmont*
897	*Redgauntlet*
898	*Sir Walter Scott*
899	*Jeanie Deans*
900	*The Fair Maid*
243	*Meg Merrilees*
244	*Madge Wildfire*
245	*Bailie Nicol Jarvie*
338	*Helen Macgregor*
339	*Ivanhoe*
340	*Lady of Avenel*
359	*Dirk Hatteraick*
360	*Guy Mannering*
361	*Vich Ian Vohr*
362	*Ravenswood*

The class was designated D29 by the LNER, 9000 added to their numbers and they were duly painted in the standard LNER lined green until the change to black with red lining in 1928. Drawings for superheating the 'Scotts' were made by NB staff in 1922 but it was not until 1925 that superheating was carried out, the first being 9895 in April 1925. 9898 and 9340 followed in November and the rest gradually until the last, 9361, in August 1936. The superheater was 192.92sq ft and total heating surface quoted as 1,346sq ft in 1926 and 1,448.62sq ft in 1938. From 1925 the LNER classified the superheated 'Scotts' as D29/2s and those still running with saturated steam boilers as D29/1. After 1936, when all were superheated, the class reverted to plain D29.

Most were fitted with tablet exchange apparatus in NBR days and hydrostatic lubricators, all except 9244, 9340, 9895 and 9898 being equipped with mechanical lubricators when superheated. The NB smokebox wingplates were removed from 1921, mostly completed before the Grouping. Last to go was 9897 in 1926. They were initially dual-brake fitted, the Westinghouse brake being removed from 1933 onwards.

The class was renumbered 2400–2415 under the LNER 1946 renumbering scheme, although the first withdrawal took place in February 1946 of 9361 which never received its new number, 2414. Twelve of the class became British Railways stock in 1948, although only five received their BR numbers, 62405 and 62410–62413. Only 62405 and 62412 were repainted in the BR lined black mixed traffic livery. The other three survivors finished in unlined black with NE on the tender (62410), lined green, lettered LNER (62411), and unlined black lettered LNER (62413). All received smokebox numberplates. The last survivor was the green painted 62411 *Lady of Avenel* withdrawn in November 1952.

A few of the names lived on – some of the locomotives withdrawn around nationalisation transferred their names to Peppercorn's new A1 Pacifics being built at the same time – *Redgauntlet, Sir Walter Scott, Meg Merrilees, Madge Widfire.*

362 *Ravenswood*, the last of the unsuperheated 'Scotts' built in December 1911, c1912. (J.M. Bentley/MLS Collections)

D29 898 *Sir Walter Scott*, wingplates removed, immediately after repainting in LNER lined green livery and before renumbering 9898, 1924. (MLS Collection)

D29/2 9362 *Ravenswood*, superheated in 1933, seen here at Haymarket shed, 10 April 1939. By this time as all the class had been superheated, they were just known as D29s. (N. Fields/MLS Collection)

Above left: **2406 (formerly** 9243) *Meg Merrilees* at Thornton Junction, 7 October 1948. (J.D. Darby/MLS Collection)

Above right: **2407 (formerly** 9244) *Madge Wildfire* on shed at St Margaret's, with V1 67605 in the background, 6 April 1947. 2407 was withdrawn in December that year. (N. Fields/ MLS Collection)

Operation

Initially the 'Scotts' were used on the main North British route expresses: Edinburgh to Carlisle, Edinburgh to Glasgow, Perth and Aberdeen. The Midland services non-stop over the Waverley route ceased during the First World War, but during and immediately afterwards they continued to share the NB main line work with the 'Superheated Scotts' and the 'Glens'.

244 *Madge Wildfire* at Aberdour on the 4.25pm Edinburgh Waverley–Aberdeen, 11 August 1910. (G.M. Shoults/ MLS Collection)

361 *Vich Ian Vohr* departing from Aberdour on the 12.10pm Edinburgh–Dundee stopping train, 5 August 1910. (G.M. Shoults/MLS Collection)

900 *The* Fair Maid with an Up troop train passing a Down coal train at Aberdour station, 21 August 1911. (G.M. Shoults/MLS Collection)

899 *Jeanie Deans* on a goods train at Aberdour, c1911. (G.M. Shoults/MLS Collection)

St Margaret's 900 *The Fair Maid* approaching Haymarket with a stopping train for Edinburgh Waverley, c1921. (MLS Collection)

At the Grouping, their allocation was:

Haymarket:	9338, 9339, 9361
St Margaret's:	9360, 9895, 9897, 9899. 9900
Dundee:	9245, 9359, 9362, 9896
Perth:	9243, 9244
Aberdeen:	9340
Hawick:	9898

After the Grouping their main line express work dwindled but they continued to haul stopping trains over the same routes. Like other 4-4-0s, they were also used for piloting heavy Edinburgh–Aberdeen expresses. The Robinson/Gresley Directors (D11/2) in

1924 and the Gresley D49s from 1927 removed the D29s from the mainline work and they were appropriately re-allocated to the following depots in the 1930s:

Haymarket: 9900
St Margaret's: 9895, 9897
Dundee: 9245, 9359, 9896
Aberdeen: 9340, 9899
Hawick: 9243, 9244
Carlisle: 9360, 9361, 9898
Thornton Jcn: 9338
Stirling: 9339
Dunfermline: 9362

A couple of runs between Edinburgh and Glasgow were logged and are in the Rail Performance Society archives, taken from Cecil J. Allen *Railway Magazine* articles of the mid-1930s. Both were on expresses, one making hard work of a near 300 ton train, the other a speedy effort with a light load.

Edinburgh Waverley–Glasgow Queen Street, mid-1930s
9339 *Ivanhoe*
10 chs, 277/295 tons

Miles	Location	Times	Speed		Gradients
0	Edinburgh Waverley	00.00			
1.2	Haymarket	03.19			L
5.5	Gogar	09.14			1/960 R
12	Winchburgh	17.07	50		1/960 R
17.6	Linlithgow	23.28	53		1/882 F
19.8	Manuel	25.46	58½		
22.3	Polmont	28.26	56		1/882 R
25.5	Falkirk (High)	32.54		1 L	
0		00.00			
3.4	Bonnybridge	06.05			L
6.3	Castlecary	09.17	54		L
8.9	Dullatur	12.04	56½		L
10.4	Croy	13.31	61		1/900 F
15.5	Lenzie	20.04	pws 30*		L
20.3	Cowlairs	26.12			1/41 F
21.9	Glasgow (Queen St)	31.02	(28½ net)	2 L	

9245 *Bailie Nicol Jarvie* with a stopping train to Hyndland, Haymarket shed in the background, c1930. (Photomatic/ MLS Collection)

The Falkirk–Glasgow scheduled time was eased by a couple of minutes later to encompass the permanent speed restriction after Croy caused by mining subsidence.

Glasgow (Queen St)–Edinburgh Waverley, mid-1930s

9897 *Redgauntlet*

5 chs, 140 tons

Lothian Coast Express

Miles	Location	Times	Speed	Gradients
0	Glasgow (Queen St)	00.00		1/41 R
1.6	Cowlairs	04.55		
3.3	Bishopbriggs	07.55		1/809 R
6.3	Lenzie	11.40	48/50	L
11.5	Croy	17.10	58	1/900 R
15.6	Castlecary	21.15	61	L
18.5	Bonnybridge	24.05	64	L
21.8	Falkirk (High)	27.40		
0		00.00		
3.2	Polmont	05.40		1/882 F
0		00.00		
4.7	Linlithgow	06.40		1/882 F
10.3	Winchburgh	12.10	61/65	1/882 R, 1/960 F
14.1	Rathbo	15.25	71	1/960 F
18.8	Saughton Jcn	19.40	66	1/960 F
21.1	Haymarket	21.45		
22.3	Edinburgh Waverley	24.24		

They continued on the same routes on stopping trains, and added Carlisle – Newcastle. Despite the existence of the D30 superheated 'Scotts', they were, once superheated, often in more demand for special working as they retained their 190lbs boiler pressure compared with the D30's 160. The two St Margaret's engines featured particularly in weekend working, such as football or seaside specials.

The class continued its main line stopping and specials work well into the era when Thompson B1s became available for this type of work.

After the Second World War, they were seen on the Fife Coast work from both Glasgow and Edinburgh and Thornton Junction became the home of many in their last days. 9895 and 9897 (as 2400 and 2402) remained at St Margaret's and worked to Galashiels, Hawick and Berwick. However, the B1 class was being multiplied rapidly and the D29s found themselves redundant and finished with 62411 on Fife Coast local work in 1952.

The NBR Class 'J' (LNER D30)

Superheating was being fitted to many locomotives in the UK at the end of the first decade of the twentieth century and the NBR directors approved its fitting to two of the 'Scott' locomotives under construction at Cowlairs at the end of 1911. The two locomotives, numbered 363 and 400, were equipped with large Schmidt superheaters and also had a number of other variations from the saturated steam 'Scotts', the main ones being the reduction in boiler pressure from 190 to 165lbs psi and the placing of the piston valves above the cylinders resulting in a higher boiler pitch. More express locomotives were required and further superheated 'Scotts' were constructed, fifteen in 1914, numbered 409–423, five in 1915, numbered 424–428 and at the end of the war, another five, 497–501, in 1920. Their key dimensions were:

Cylinders (2-inside):	20 x 26in
Coupled wheel diameter:	6ft 6in
Bogie wheel diameter:	3ft 6in
Boiler pressure:	165lbs psi

Stephenson notion with 8" piston valves (363 & 400) and 10in piston valves (the rest)
Heating surface: 1,573sq ft (incl 266.4sq ft superheating) (363 & 400)

	1,640.7sq ft (incl 355.2sq ft superheating) (409–428 & 497–501)
Grate area:	21.13sq ft
Axleload:	19 tons 14 cwt (19 ton 8 cwt 409–428 & 497–501)
Weight (Engine):	57 tons 6 cwt (57 tons 16 cwt (409–428 & 497–501)
(Tender):	46 tons (46 tons 13 cwt 409–428 & 497–501)
(Total):	103 tons 6 cwt (104 tons 9 cwt 409–428 & 497–501)
Water capacity:	4,235 gallons
Coal capacity:	7 tons
Tractive effort:	18,700lbs

The locomotives were named:

363	Hal 'o the Wynd
400	The Dougal Cratur
409	The Pirate
410	Meg Dods
411	Dominie Sampson
412	Laird o' Monkbarns
413	Caleb Balderstone
414	Dugald Dalgetty
415	Claverhoues
416	Ellengowan
417	Cuddie Headrigg
418	Dumbiedykes
419	The Talisman
420	The Abbot
421	Jingling Geordie
422	Kenilworth
423	Quentin Durward
424	Lady Rowena
425	Kettledrummle
426	Norna
427	Lord Glenvarloch
428	Adam Woodcock
497	Peter Poundtext
498	Father Ambrose
499	Wandering Willie
500	Black Duncan
501	Simon Glover

The twenty-five later locomotives were equipped with the Robinson superheater rather than the Schmidt and had 10in piston valves instead of 8in. At the Grouping, the two 1912 locomotives were classified as D30/1, and the other twenty-five as D30/2. Like all NB engines, they had 9000 added to their numbers and were initially painted in the LNER lined green livery, this changing to black with red lining in 1928. Boiler changes took place and the boilers with Schmidt superheaters on the D30/1s

The Pioneer
'Superheated Scott', 400 *The Dougal Cratur*, as built in September 1912 with smokebox wingplates and Westinghouse brake.
(MLS Collection)

Superheated 'Scott'
411 *Dominie Sampson*, as built in 1914.
(F. Moore/MLS Collection)

were replaced with Robinson superheater boilers – 9400 in 1927 and 9363 in 1936. The number of the superheating elements was reduced in the 1930s to 192.92sq ft and the total heating surface reduced to 1,448.62sq ft.

Tablet exchange apparatus was fixed to the tender of eleven D30/2s in late NB or early LNER days but as some tenders were exchanged, the locomotives so fitted would change. At the end of their careers, four, then numbered 62418, 62438–62440, were still equipped. Removal of smokebox wingplates took place between 1921 and 1925. Mechanical lubricators were standard, apart from 410 which was fitted with Detroit hydrostatic lubricator in 1917 and retained it until withdrawal. The locomotives were dual-braked until 1935-37 when the Westinghouse air-brake equipment was removed.

The locomotives were renumbered in 1946 as 2417–2442 in building order. 9400, allocated 2416 was withdrawn in June 1945 and never received the number. All then received BR numbers in 1948, becoming 62417–62442 apart from 2433 *Lady Rowena* which was withdrawn in November 1947. The other D30/1, 62417 *Hal o' the Wynd* was withdrawn in January 1951, its name then being applied to new A1 60116, (and many years later to Pendolino 390.131). The rest remained in service until the late 1950s, the withdrawal of 62430 *Jingling Geordie* starting a steady reduction in the class until just two were left in 1960 – 62421 *Laird o' Monkbarns* and 62426 *Cuddie Headrigg*. Both were withdrawn in June that year.

D30/2 9416 *Ellengowan* with Westinghouse brake and mechanical lubricator clearly visible in this photograph, c1935. (Real Photographs/MLS Collection)

Superheated 'Scott'
D30/2 9426 *Norna* in LNER apple green livery, September 1925. (MLS Collection)

Above left: D30/2 2431 *Kenilworth* at Thornton Junction, 6 April 1947. (J.D. Darby/MLA Collection)

Above right: D30 62441 *Black Duncan* in early British Railways lined black mixed traffic livery, at Dunfermline shed, 31 August 1952. (A.C. Gilbert/MLS Collection)

D30 62437 *Adam Woodcock* ex-works at Haymarket, 21 October 1951. (MLS Collection)

Operation

Between 1912 and 1914, 363 and 400 were based at St Margaret's and were used on the non-stop Waverley route trains, the London–Edinburgh via the Midland Railway. These ceased at the beginning of the First World War, and they then joined the 1914 built engines on express work to Glasgow and Perth as well as other trains to Carlisle. 409 was tried on the West Highland line in the summer of 1914, but they stayed mostly working from Carlisle, Edinburgh and Glasgow during the war. One regular turn was the naval 'Jellicoe' trains worked by a couple of superheated 'Scotts' from Carlisle to Perth.

Between the end of the First World War and the Grouping, they were the most frequent performers on the Glasgow–Haymarket non-stop expresses scheduled in the even hour and often quite substantially loaded. Three logs are tabled overleaf of 1914 built engines on the westbound run and one on the eastbound.

62429 *The Abbot* at Thornton Junction, 4 August 1957.
(MLS Collection)

The first 'Superheated Scott' 400 *The Dougal Cratur*, working the Carlisle–Edinburgh non-stop Waverley route express with Midland coaches through from St Pancras and Leeds, c1913.
(MLS Collection)

Haymarket–Glasgow Queen Street

		416 *Ellengowan* 280 tons		414 *Dugald Dalgetty* 310 tons		412 *Laird o' Monkbarns* 360 tons		
Miles	Location	Times	Speed	Times	Speed	Times	Speed	Gradients
0	Haymarket	00.00		00.00		00.00		
7	Ratho	10.31	55	10.49	58	11.20		1/960 R
16.4	Linlithgow	20.19	61	21.22	54/64	22.18	52	
21.1	Polmont	25.28	55½	27.01	50	28.15	47	1/882 R
24.3	Falkirk	29.58	pws	30.14	57	31.31	56	L
30.6	Castlecary	36.27	58	36.24	61	38.10	57	L
39.9	Lenzie	46.41	55/64	46.12	57½/62	47.42	59/63	1/900 F
44.6	Cowlairs	53.51	sigs	51.46	sigs	52.47		L, 1/41 F
46.1	Glasgow Queen St	58.54	1 E	58.57	1 E	58.50	1 E	
		(57 net)		(58 ½ net)				

Glasgow Queen Street–Haymarket

413 *Caleb Balderstone* 344/370 tons

Miles	Location	Times	Speed		Gradients
0	Glasgow Queen St	00.00		(banked)	1/41 R
1.5	Cowlairs	05.45			
6.2	Lenzie	12.57	46		L
15.5	Castlecary	23.46	53		L
21.8	Falkirk	30.27	61		L
25	Polmont	33.50	59		L
29.7	Linlithgow	38.45	62		1/882 F
35.3	Winchburgh	44.58	54		1/882 R
41.8	Gogar	52.17	easy		1/960 F
46.1	Haymarket	57.34		1½ E	

At the Grouping they were allocated as follows:

Eastfield: 9410, 9411, 9497, 9499, 9500
Haymarket: 9413–9416, 9424, 9428
St Margaret's: 9363, 9400, 9417, 9423, 9426
Carlisle: 9412
Perth: 9409, 9418
Dundee: 9419–9422, 9425, 9427
Thornton Jn: 9498, 9501

The Eastfield engines worked to Edinburgh on expresses including 'The Queen of Scots' Pullman train and 'The Lothian Coast Express', the expresses between the two cities being tightly timed and often loaded up to 350 tons and more. They also worked to Stirling and Perth, the Fife Coast, the latter with Dundee and Thornton engines reciprocating. The D30s were competent machines and were noted for their starts up the 1 in 41 Cowlairs Bank on the Glasgow starters. O.S. Nock had a number of runs in the 1920s on the Aberdeen–Edinburgh run with very heavy loads, especially on the sleeper services which were double-headed before the arrival of the P2 2-8-2s. On one occasion, a Reid Atlantic was piloted by superheated 'Scott' 9413 of Haymarket. Nock reported that both engines appeared to working 'flat out' on station departures and on the steepest gradients.

Aberdeen–Dundee, Easter Monday 1934

9869 *Bonnie Dundee* **(Atlantic) & 9413** *Caleb Balderstone* **(D30/2)**

523/565 tons

7.35pm Aberdeen–King's Cross sleeper

Miles	Location	Times	Speed	Gradients
0	Aberdeen	00.00		
4.8	Cove Bay	09.33	41/35	1/154 R, 1/102 R
7.1	MP 234	13.11	41	1/164 R
11.6	Muchalls	18.05	64½	1/126, 1/310 F
<u>16.2</u>	<u>Stonehaven</u>	<u>23.42</u>		
0		00.00		
2.6	Dunnottar Box	06.39	38½	1/92 R
5.5	Carmont	11.13	34½	1/102 R
11.2	Fordoun	17.20	62½	1/170 F
14.5	Laurencekirk	20.41	52	1/285 R
19.8	Craigo	26.23	62½	1/104 F
21.9	Kinnaber Junction	29.10		
<u>24.5</u>	<u>Montrose</u>	<u>33.40</u>		
0		00.00		
2.1	Usan Box	04.53	32½	1/88 ½ R
4.9	Lunan Bay	09.28	38½	1/111 R
7.5	Inverkeilor	12.11	pws 45*	1/100 F
10.7	Letham Grange	16.54	62½	L
<u>13.7</u>	<u>Arbroath</u>	<u>21.15</u>		
0		00.00		
1.8	Elliott Junction	03.27		
6.1	Carnoustie	08.25	60	L
13	Broughty Ferry	14.59	64½	L
<u>17</u>	<u>Dundee</u>	<u>21.05</u>		

Nock timed another run on the continuation south with Dundee's D30/2 9427 *Lord Glenvarloch* piloting Reid Atlantic 9877 and Cecil J. Allen logged the same 'Scott' piloting Atlantic 9510 on the Aberdeen–King's Cross sleeper. Again, both recorders commented particularly on the noise made by the pairs of engines being worked very hard.

		Dundee–Edinburgh Waverley, earl 1930s					
		9877 *Liddersdale*		9510 *The Lord Provost*			
		9427 *Lord Glenvarloch*		9427 *Lord Glenvarloch*			
		434/475 tons		511/550 tons			
Miles	Location	Times	Speed	Times	Speed	Gradients	
0	Dundee	00.00		00.00			
2.7	Tay Bridge Box	07.05		07.05		1/74 R	
4.6	St Fort	09.40	65	09.45	57	1/100 F	
8.3	Leuchars Junction	13.40	62	13.38	63	L, 1/100 F	
11.6	Dairsie	17.40	45	17.22	50½	1/160 R	
14.6	Cupar	21.00	57½	20.27	61	L	
16.9	Springfield	23.40	48	23.12	47½	1/161 R	
20.1	Ladybank Junction	27.35	60	26.52	58½	L	
24.3	Lochmuir Box	34.05	33	32.30	35	1/85, 1/105 R	
25.9	Markinch	35.35	60	34.40	60/pws	1/102 F	
28.5	Thornton Junction	38.50		38.32	20*		
	MP 29	-	35½	-	32	1/131, 1/158 R	
33.3	Kirkcaldy	45.45	64½	46.19	63½	1/100, 1/143 F	
39	Burntisland	52.20	35*	53.20	20*		
41.8	Aberdour	56.25	48/37½	57.48	45½/39	1/100 R	
43.2	Dalgetty Box	58.35	39	59.50	39	1/100 R	
46	Inverkeithing	62.05	60/45*	63.27	54/40*	1/94 ½ F	
48	Forth Bridge North	65.45	28	67.50	23½	1/70 R	
52.7	Turnhouse	72.10	65	74.42	63½	1/100 F	
58	Haymarket	78.20	sig stop	80.35			
<u>59.2</u>	<u>Edinburgh Waverley</u>	<u>82.40</u>		<u>83.55</u>			

However, by 1930 not only were the Scottish Director D11/2s on the scene, but also the D49 4-4-0s and the Gresley Pacifics, especially on the Edinburgh–Glasgow and Aberdeen turns, so the D30s were dispersed as follows:

Eastfield:	9409, 9497, 9498	Dundee:	9413, 9420, 9425, 9427
Haymarket:	9411, 9412, 9415, 9416, 9424, 9428	Thornton Jn:	9410, 9421, 9422, 9500, 9501
St Margaret's:	9363, 9400		
Carlisle:	9426	Hawick:	9419, 9499
Perth:	9417, 9418	Aberdeen:	9423

O.S. Nock recorded a Haymarket D30 on an Edinburgh–Glasgow semi-fast train in 1936 which was quite smart and typical of D30 running on that route at that time. Another D30 was recorded on the 3.48pm Perth–Edinburgh. The logs are shown below and overleaf.

Edinburgh Waverley–Glasgow Queen Street, 1936
1.20pm King's Cross–Glasgow
9415 *Claverhouse*
232/245 tons

Miles	Location	Times	Speed	Gradients
0	Edinburgh Waverley	00.00		
1.2	Haymarket	03.10		
3.5	Saughton Junction	06.07	53½	1/960 R
8.2	Ratho	11.19	65	1/960 R
14.2	Philipstoun	18.48	pws 42*/62	1/882 F
17.6	Linlithgow	22.47		
0		00.00		
2.2	Manuel	04.03	53/ pws	
4.7	Polmont	08.08		
0		00.00		
3.2	Falkirk (High)	06.13		
0		00.00		
6.2	Castlecary	08.23	61½	L
10.3	Croy	12.23	60	L
	Waterside Jcn	-	64½	1/900 F
15.5	Lenzie	18.55	pws 30*/52	L
20.2	Cowlairs	25.06		1/41 F
21.8	Glasgow Queen St	31.08		

Perth – Dunfermline, 26.8.1938
9417 *Cuddie Headrigg*
7 chs, 213 tons

Miles	Location	Times	Speed		Gradients
0	Perth	00.00		¼ L	
2	Hilton Junction	05.40			
3.5	Bridge of Earn	07.34	52½	1¾ L	L
	MP 44	-	54		L
	MP 43	-	45		1/75 R
	MP 42	-	38½		1/75 R
	MP 41	-	30½		1/75 R
	MP 40	-	29		1/75 R

		Perth – Dunfermline, 26.8.1938 9417 *Cuddie Headrigg* 7 chs, 213 tons			
Miles	Location	Times	Speed		Gradients
	MP 39	-	27½		1/74 ½ R
	MP 38	-	26½		1/74 ½ R
10.2	Glenfarg	19.44		T	
13.6	Mawcarse Junction	23.45	61		1/94 F
17.2	Kinross	27.46	52	1 L	1/198 R
18.1	Lochleven	-	60		1/154 F
	Ketty	-	45* (colliery slack)		
25.3	Cowdenbeath	37.32	40/10*	1¼ E	1/80 R
30.4	Dunfermline	49.13	sigs	2 L	

The train continued. dogged by signal checks including a dead stand at Inverkeithing, and arrived at Edinburgh Waverley 8½ minutes late. On 18 September 1936, Gerald Aston had logged a Pacific from Aberdeen to Dalmeny where he changed into a Fife Coast–Glasgow express hauled by a 'Scott', 9501 *Simon Glover*. It had a train of seven coaches, 193/200 tons and ran the 17.7 miles to Falkirk in 21 minutes exactly, with a top speed of 66mph at Maneil. It carried on in similar style, had reached 60mph by Greenhill Upper Junction (4½ miles) and touched 66 again between Croy and Lenzie, passing Cowlairs 19.9 miles in 22 minutes six seconds, and with a cautious descent down Cowlairs Bank, completed the run to Glasgow Queen Street in 29 minutes 12 seconds.

As well as working semi-fast expresses and stopping trains from these depots they were often used as pilots on heavy East Coast trains both north and south of Edinburgh. The Carlisle engine was also used on the former North Eastern route to Newcastle as well as stopping trains to Hawick and Galashiels with the Hawick engines. Occasionally they would find themselves on express work after the failure of other larger locomotives and one Manchester Locomotive Society member, H.D. Bowtell, recounted travelling on a 14-coach 442 ton train Dundee–Aberdeen train double-header by D11/2 6401 and 9413 *Caleb Balderstone*. The Director developed a hot box and had to be removed at Stonehaven and the D30 had to continue to Aberdeen alone over the cliffs above Cove with the heavy load which it managed successfully.

Below left: D30 9417 *Cuddie Headrigg* pilots D49 246 *Morayshire* with what appears to be a train bound for Carlisle and the Midland main line, with two Midland coaches at the front, c1930. (Colling Turner/MLS Collection)

Below right: St Margaret's D30/1 9363 *Hal o' the Wynd* departing from Waverley station through Princes Street Gardens, 4 August 1938. (L. Hanson/MLS Collection)

As late as 1937 a pair of Reid 4-4-0s were discovered in charge of 'The Flying Scotsman' on its route north of Dundee. R.E, Charlewood was in the train and recorded the log below:

Dundee–Aberdeen, 30.8.1937
9737 (D31) + 9423 *Quentin Durward*
10 chs, 350 tons
'The Flying Scotsman'

Miles	Location	Times	Speed	Gradients
0	Dundee	00.00		
4	Broughty Ferry	07.13		L
6.4	Monifieth	09.55	54	L
10.9	Carnoustie	14.20	62	L
17	Arbroath	20.34		1½ E
0		00.00	T	
3	Letham Grange	06.32		1/103½ R
8.6	Lunan Bay	12.43		1/100 R
9.7	MP 26 ¾ (summit)	14.06	47	1/93 R
11.6	Usan	16.20	60/ 40*	1/111 F
13.1	Montrose	18.47		1¼ E
0		00.00	T	
1.6	Kinnaber Junction	05.10		1/90 R
6.8	Marykirk	10.37	57	L
8.9	MP 209 ¼	13.31	31	1/104, 106 R
11.2	Laurencekirk	15.08		
14.7	Fordoun	18.41	60	L
18.7	Drumlithie	23.21		1/170 R
	MP 218 ¼ (summit)	24.20	48	1/141 R
26	Stonehaven	33.06	3 E	
0		00.00	1 E	
2.5	MP 227 ½	05.05	pws 20*	1/100 R
4.5	Muchalls	07.25	60	1/163 F
8	Portlethen	11.45	48	1/126 R
9	MP 234 (summit)	12.53	52	1/97 R
11.25	Cove	15.09	62	1/102 F
14	MP 239	17.42	67	1/118 F
15.5	Ferryhill	19.45		
16.1	Aberdeen	21.19	(19½ net)	¾ E

During the Second World War, more D30s were relegated to the smaller depots, eight being at Hawick for working to Carlisle, Newcastle as well as Edinburgh and nine were at Thornton and Dunfermline. The two D30/1s were still at St Margaret's but 9400 was the first withdrawal in June 1945 and one D30/2 was withdrawn in the last month before nationalisation – 9424 *Lady Rowena*. The other D30/1, 9363 as 62417, was withdrawn in January 1951 but the rest carried on working mainly stopping trains until the late 1950s. Their distribution in the 1950s was:

	1950	1957
St Margaret's:	62421, 62424	62421, 62424
Haymarket:	62437	62437
Thornton Jn:	62418, 62419, 62429-62431, 62442	62418, 62419, 62429, 62431, 62442
Hawick:	62420, 62422, 62423, 62425, 62428, 62432, 62440	62420, 62422, 62423, 62425, 62428, 62432, 62435, 62440
Stirling:	62426	62426
Dundee:	62427, 62434, 62436, 62438	62434, 62438
Bathgate:	62439	62439
Dunfermline:	62441	62427, 62436, 62441

D30 62436 *Lord Glenvarloch* pilots D49 62733 *Northumberland* with the 4.15pm Perth–Edinburgh approaching Glenfarg, 23 July 1949. (J.D. Darby/MLS Collection)

From this it will be noted that very little movement between sheds took place in the 1950s and the 1957 allocations remained static during the final years of their operation. The final two were 62421 withdrawn from St Margaret's and 62426 from Stirling. Their work was modest in the 1950s, but I have discovered one log in the Rail Performance Society's archive from 1952.

Stirling–Edinburgh, 27.8.1952

62426 *Cuddie Headrigg*

5 chs, 140 tons

Miles	Location	Times	Speed		Gradients
0	Stirling	00.00		¼ L	L
4.1	Plean	09.27	pws 5*/ 36½		1/118 R
7.8	Larbert	14.26	58	2½ L	1/126 F
0		00.00		1½ L	
1.4	Carmuirs East Jn	02.29	44		L
3	Grahamston	05.38		1½ L	
0		00.00		1½ L	
3.3	Polmont	06.49	34/31*		1/100 R
8	Linlithgow	13.06	51/44/46		1/882 F
13.6	Winchburgh	18.59	55/35*		1/960 F
17.6	Ratho	24.39	56		1/960 F
20.2	Gogar	-	66½		1/960 F
23.8	Haymarket West Jn	30.44	34*/ sig stand		
25	Edinburgh Princes St	35.19	(34 net)	¾ L	

Below left: D30 9424 *Lady Rowena* leaving Waverley with a northbound local service, 22 April 1946. 9424 would be the first D30/2 to be withdrawn just over a year later. (MLS Collection)

Below right: Hawick's 62425 *Ellengowan* with a stopping train to Galashiels, at Hawick, 7 August 1949. (R.D. Pollard/MLS Collection)

Dundee's 62434 *Kettledrummle* prepares to stop at Dundee Esplanade, c1957.
(MLS Collection)

The NBR Class M (LNER D31)

In 1884, Holmes built six 6ft 6in 4-4-0s based on earlier proposed Drummond engines, smaller than the 'Abbotsford' class but with a boiler that was standard with his 0-6-0 goods engines. They were numbered 574–579 and were known as the '574 class' rather than their official designation 'M'. They were built for the Glasgow–Edinburgh trains rather than the longer distance expresses. Their dimensions were:

Cylinders (2-inside):	17 x 26in
Coupled wheel diameter:	6ft 6in
Bogie wheel diameter:	3ft 6in
Boiler pressure:	140lbs psi

Stephenson motion with slide valves

Heating surface:	1,059sq ft
Grate area:	17sq ft
Axleload:	15 tons 9 cwt
Weight (Engine):	43 tons 10 cwt
(Tender):	32 tons
(Total):	75 tons 10 cwt
Water capacity:	2,500 gallons
Coal capacity:	4 tons
Tractive effort:	11.464lbs

A development of the class with larger cylinders and slightly larger boilers and a 9in longer wheelbase was built in 1890, and numbered twelve locomotives, 633-642, 36 and 37, which were known as the '633' class. Their changed dimensions were:

Cylinders (2-inside):	18 x 26in
Boiler pressure:	140lbs psi (1894/5 locos had 150lbs psi)
Heating surface:	1,262sq ft
Grate area:	20sq ft
Axleload:	15 tons 12 cwt
Weight (Engine):	46 tons
Tractive effort:	13,770lbs

These engines were needed for the Edinburgh–Perth and Edinburgh Aberdeen services introduced after the building of the Forth Bridge. With the increase in loads and services more were required and six more were constructed in 1894, Nos. 262, 293, 312, 404, 211 and 212. Another six, 213–218 were built in 1895 for the Waverley route and the Midland Company trains for Scottish destinations.

In 1894, the previous agreement for North Eastern engines to work the Newcastle East Coast expresses through to Edinburgh was terminated and the North British took over services north of Berwick. This led to the construction of similar locomotives but with slightly larger cylinders and boiler. Larger tenders were required for the new East Coast trains. Twelve were constructed in 1898, numbered 729-740 and six in 1899 numbered 765-770. These engines became known as the '729' class although all these engines from 1890 onwards were officially designated as class 'M'. The changed dimensions of the '729' class from the '633' class were:

Cylinders (2-inside):	18¼ x 26in
Boiler pressure:	175lbs psi
Heating surface:	1,350sq ft
Axleload:	16 tons 10 cwt
Weight (Engine):	47 tons 5 cwt
(Tender):	39 tons
(Total):	86 tons 5 cwt
Water capacity:	3,500 gallons
Coal capacity:	6 tons
Tractive effort:	16,514lbs psi

In 1911, the members of the '574' class were rebuilt with Reid boilers with the dimensions of the '729' class boilers and as boilers of the '633' and '729' boilers were due for renewal between 1918 and 1922, they were rebuilt to the Reid design used for the rebuilding of the '574s'. This slightly increased the weight of the engine to 46 tons 8 cwt and the axleload to 17 tons 4 cwt, but reduced the overall heating surface to 1,266sq ft. During the Reid years the Holmes curved cabs were replaced with a single side-window cab. At the Grouping, all forty-eight had been rebuilt (see Appendix for date of conversion). The LNER classified all the locomotives as D31 and in 1927 subdivided the class, thirty with the Reid design as D31/1 and eighteen with variations by Chalmers in 1921/2 as D31/2. Basic difference was in the springing, but these differences were soon removed and the sub-division fell into disuse. The engines were renumbered at the Grouping with the addition of the standard 9000 applied to North British locomotives. The locomotives lost their smokebox wing plates, the Chalmers rebuilds in 1921/2 and the rest in the early LNER years.

The first withdrawal took place in 1931 and more withdrawals took place between 1933 and 1939, but seventeen remained at the beginning of the Second World War, during which only one was condemned. The remaining sixteen were given the numbers 2059-2074 in the LNER 1946 scheme and seven of these came into British Railways stock in 1948 with the added 60,000. However, when more of the Thompson K1 2-6-0s were being built and required more of the 62000 series numbers, the four D31s remaining in 1949 were again renumbered 62281-62284.

The pioneer of the '633' class, 633 itself, as built in 1890, seen here c1900. (MLS Collection)

62065 had however just been withdrawn so just 62059, 62060 and 62072 became 62281-62283. They were all in plain black livery, with BRITISH RAILWAYS on the tender, but 62281, the last survivor, received the full BR lettering and a smokebox door numberplate. 62060, as 62282 was the only one to acquire lined black mixed traffic livery and 62283 also got a smokebox door numberplate. 62065 was withdrawn in 1949, 62282 in 1950, 62283 in 1951 and 62281 in December 1952.

Above left: **633 again,** now as rebuilt in 1918 with the Reid boiler and cab, but still retaining its wingplates, at Balloch, c1920. (Photomatic/ MLS Collection)

Above right: **732** was rebuilt in 1922 and is seen here in 1923 at Inverkeithing. Note Reid cab and the removal of the wingplates from engines rebuilt by Chalmers. (T.G. Hepgurn/ Rail Archive Stephenson)

9734 in the LNER black livery applied from 1928, seen at Kittybrewster, 10 August 1937. (MLS Collection)

9635 at Carlisle on a Silloth local train on 8 April 1946. Plain black livery with NE on tender, just before renumbering as 2059 (and ultimately the last survivor as BR 62281). (H.C. Casserley/MLS Collection)

Below left: **2073, the** former 769, dumped at Inverurie after its withdrawal in December 12947, seen here on 27 March 1948. (F.R M. Fysh/MLS Collection)

Below right: **The only** BR survivor, 62281, four months before its withdrawal and class extinction, on Carlisle Canal shed, in the company of Carlisle allocated A3, 60095 *Flamingo*, 30 August 1952. (MLS Collection)

Operations

As stated earlier, the '574s' were built for the Edinburgh – Glasgow expresses rather then the heavier Waverley rout trains for which the 'Abbotsfords' were still provided. The '633s' worked the new services over the Forth Bridge and augmented the Drummond engines on the Waverley route. The '633s' were immediately involved in the Edinburgh–Berwick services after the completion of the NBR/NER agreement in April 1894 and were involved in the East/West Coast races to Aberdeen in 1895, the NER engines working to Edinburgh and the '633s' on to Aberdeen. The '633s' after 1894 worked the Newcastle–Edinburgh services that stopped at Berwick from that town northwards.

The fastest time in the 1895 races north of Edinburgh was with 1890

The last of the '633' class, 218, built in 1895, at Silloth with a special outing run for Carr's Flour Mill staff, 1905. (MLS Collection)

633 on the 6.52pm Perth–Edinburgh near Loch Leven, 14 August 1911. (G.M. Shoults/MLS Collection)

767 approaching Aberdour with a stopping train for Dundee, 1911. (G.M. Shoults/MLS Collection)

216, a '633' engine built in 1895 and rebuilt in November 1920, on a Silloth to Carlisle train, c1921. (Real Photographs, MLS Collection)

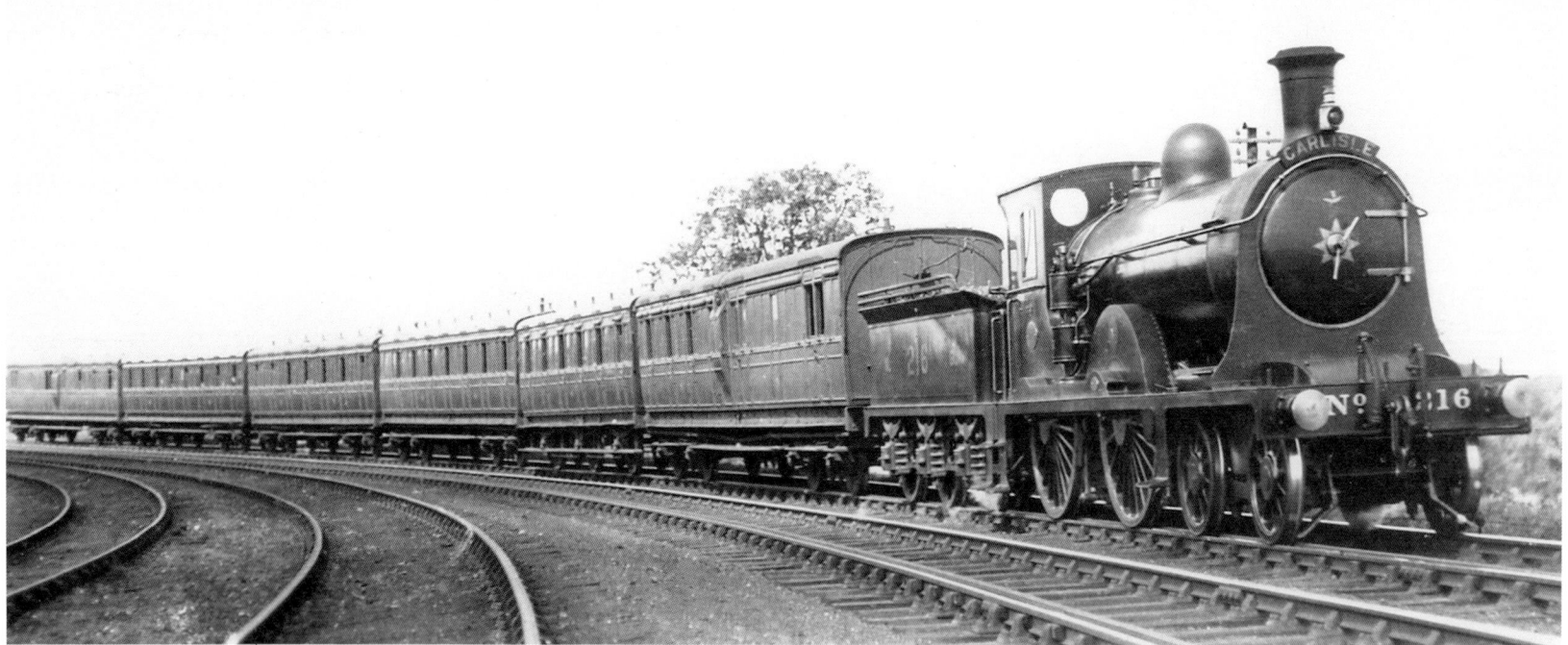

built 293 which ran the 59.2 miles to Dundee in 59 minutes exactly and 262 which covered the 71.3 miles on to Aberdeen in 77 minutes. The train was a lightweight 3-coach 86 tons only. The '729' class appeared in 1898 too late to participate in the 'races' but covered the Edinburgh–Berwick sections of the East Coast work until 1904 when a new agreement permitted the NER engines to resume working through to Edinburgh. All forty-eight engines around the turn of the century were based at Cowlairs, Haymarket, St Margaret's, Perth, Aberdeen and Carlisle. With larger engines available from 1903, these three classes cascaded to semi-fast and stopping trains.

After rebuilding in the immediate pre-Grouping years the engines were allocated as follows:

Carlisle:	9036, 9216, 9218
Hawick:	9404, 9576, 9633
St Margaret's:	9213, 9732, 9737, 9740
Haymarket:	9759, 9729, 9730, 9733 – 9736
Berwick:	9766
Blaydon:	9312
Eastfield:	9217, 9293, 9768
Parkhead:	9215
Stirling:	9037, 9211, 9577
Perth:	9214, 9574, 9634, 9637, 9731, 9738
Dunfermline:	9262, 9640 – 9642, 9769
Bathgate:	9578, 9639
Thornton Jn:	9212, 9635, 9636, 9638, 9767, 9770
Dundee:	9765
Aberdeen:	9575, 9739.

Berwick shed closed in 1924, 9766 moving initially to Tweedmouth, then joining 9312 at Blaydon for work in the Border country to Riccarton Junction. The Carlisle engines worked to Silloth and Hawick and the Dunfermline engines worked to the Fife Coast, the remainder running stopping services over the main lines radiating from Edinburgh. One of their frequent duties in the 1920s and early 1930s was piloting heavy trains including the sleeper services over the Edinburgh–Aberdeen main line.

O.S. Nock recorded one run when a D31 9769 piloted a Reid Atlantic, 9902 *Highland Chief,* on the 7.35pm Aberdeen–King's Cross sleeper. The pair had a gross load of 470 tons and achieved 39mph on the 1in 102 past Cove Bay, falling to 35 at the summit before Portlethen, covering the 16.2 miles to the Stonehaven stop in 24 minutes. The climb out of Stonehaven was first class with 42 reached the short level stretch near MP 111½ dropping to only 37mph on the 1 in 102 to Carmont. The pair then reached 68mph down the 1 in 170 to Fordoun, fell to 49 at Laurencekirk and the reached almost 70mph on the 1 in 104 descent to Craigo. The 24½ miles from Stonehaven to Montrose took exactly 32 minutes comfortably within the schedule. Another effort on the start up the 1 in 88/111 to Lunan Bay produced 30mph on the steepest stretch rising to 34½ further up the bank and another 68 through Inverkeilor saw the sleeper into Arbroath on time. The last level stretch into Dundee produced nothing over 60mph but that was enough for timekeeping, the final 17 miles being completed in 22½ minutes. 9737 piloted D30

9634, which was sent to the former Great North of Scotland lines in 1925, double-heads a D40 with a heavy stopping train that has arrived from the north, c1930. 9634 still bears the LNER lined green livery. (H. Gordon Tidey/MLS Collection)

The last survivor, 62281, arriving at Carlisle Citadel station with a local train from Silloth, 18 August 1950. (MLS Collection)

Not only had passenger traffic grown, the North British Railway served the coasts of Scotland and the fishing industry and the countryside where beef farming was prominent. The growth of fitted freight traffic requiring faster runs to the meat and fish markets had been met initially by equipping some 0-6-0s with vacuum brakes, but in 1905 the directors saw a need for something more appropriate and Reid designed and constructed the duly authorised 4-4-0s with 6ft coupled wheels – twelve locomotives that became known as the 'Intermediates' (the term of the time for what later were designated as 'mixed traffic' locomotives). The dozen locomotives were numbered 882–893 and were delivered from Cowlairs Works in 1906 and 1907. Their key dimensions were:

Cylinders (2-inside):	19 x 26in
Coupled wheel diameter:	6ft 0in
Bogie wheel diameter:	3ft 6in
Boiler pressure:	190lbs psi

Stephenson motion with piston valves

Heating surface:	1,760sq ft
Grate area:	22.5sq ft
Axleload:	18 tons 18cwt
Weight (Engine):	53 tons
(Tender):	40 tons
(Total):	93 tons
Water capacity:	3,525 gallons
Coal capacity:	7 tons
Tractive effort:	21,053lbs

9423 on 'The Flying Scotsman' from Dundee to Aberdeen on a load of 340 tons in August 1937, tabled earlier (see page 90).

In 1925, redundant Great Northern D1s were sent to Scotland enabling some D31s with lighter axleload to replace withdrawn Great North of Scotland 4-4-0s. More gradually moved to the north and by 1935 the following were running on routes of the former GNoS – 9037, 9211, 9212, 9215, 9404, 9575, 9634, 9730, 9733-9740. A couple returned to former NB territory but in the 1940s a swath of further D31s found their way north – 9642, 9731, 9732 and 9765-9770. The following returning to the North British lines to end their days there – 9215, 9642, 9729, 9735. The D31s were allocated to Kittybrewster, Keith and Elgin sheds and worked trains on the main Aberdeen – Elgin and Deside lines plus braches to Lossiemouth, Buchan and Fraserburgh. They were genuine mixed traffic engines working fish and goods trains as well as local passenger services. The Southern area D31s remained at Carlisle working to Newcastle as well as Silloth. The St Margaret's engines helped with heavy wartime goods traffic and banking duties at Niddrie and Portobello.

The last survivors were 62283 at Bathgate and 62281 and 62282 at Carlisle.

The NBR Class 'K' (LNER D32)

We move back to the time of William Reid at Cowlairs, the first decade of the twentieth century.

They worked very successfully on the traffic for which they were designed and were true mixed traffic locomotives. They all were taken into LNER stock at the Grouping and were superheated when reboilered between 1923 and 1926. The LNER classified the saturated locomotives as D32/1 and the superheated engines as D32/2, dropping the subdivision when all had been superheated. The new boilers had a lower boiler pressure of 180lbs psi, heating surface of 1,346.06sq ft and grate area of 21.13sq ft, but superheat surface of 192.92sq ft. Tractive effort was reduced to 19,945lbs and engine weight and axleload increased slightly. Other boilers used on the class later had minor changes in total heating surface, but no other dimension differences.

The D32s were renumbered 9882–9893 and lost their smokebox wingplates between 1921 and 1924. When superheated, all bar two were equipped with Wakefield mechanical lubricators, 9884 and 9891 getting Detroit four-feed hydrostatic lubrication instead. 9886 and 9889 required new frames in 1925 and 1929 respectively and three, allocated to the former North Eastern lines in the 1930s were equipped with the Raven cab signalling apparatus. (9882, 9887 and 9888). All were dual-brake fitted but the Westinghouse system was removed in the mid-1930s, the locomotives getting steam brakes on engine and tender retaining vacuum brakes for the train.

At first the LNER turned them out in black with red lining, then had second thoughts and liveried them in lined green from the end of 1925, only to return them to black again from 1928. They all survived the Second World War and were renumbered 2443–2454 in the LNER

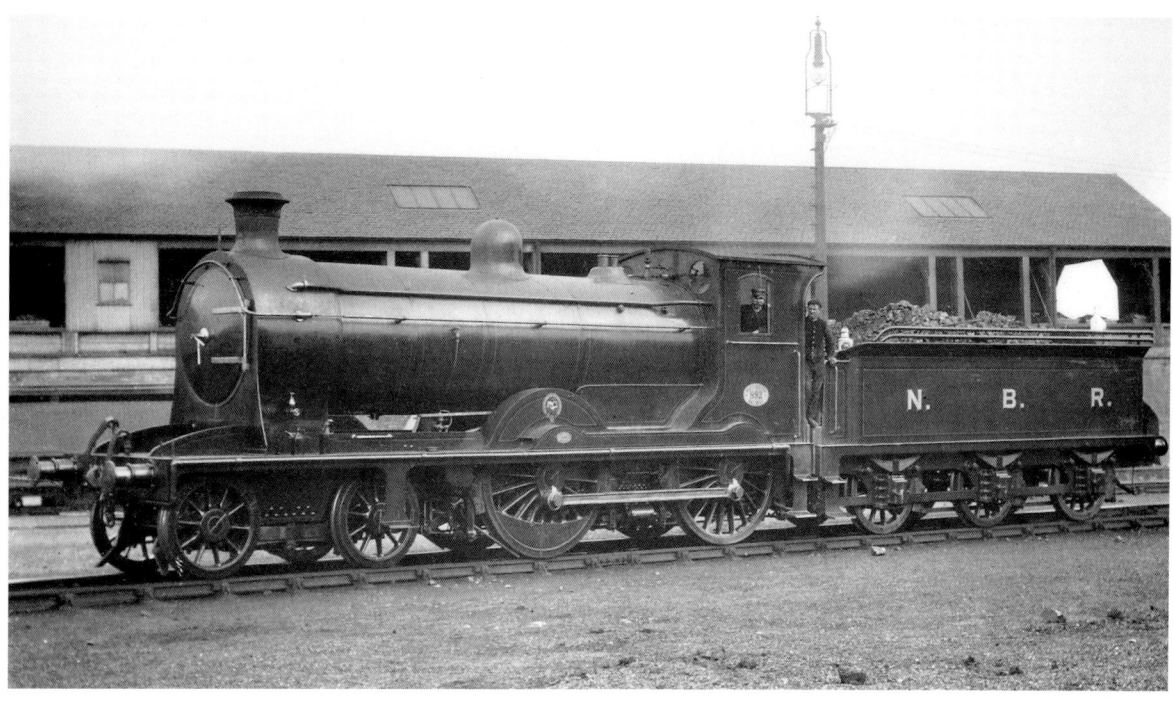

Above: **892 in** NBR livery as built in 1907. (MLS Collection)

Below: **9893 built** in December 1895 and rebuilt with new boiler at Inverurie in 1920, seen c1929. (Photomatic/MLS Collection)

1946 scheme. 2447 and 2452 were withdrawn at the end of 1947 but the other ten all entered British Railways stock in 1948, being allocated 62443–62454, though only 62451 bore the new BR number. All bar this locomotive were withdrawn in 1948 or 1949. 62451 was withdrawn in March 1951.

Operations

Most of the initial allocation went to Glasgow Eastfield, though two went to Berwick and one to Montrose for long distance freight traffic. Some of the Eastfield engines moved later to St Margaret's for working to Newcastle and Carlisle. As well as freight work, the Eastfield engines worked passenger trains on the West Highland line and the St Margaret's similarly passenger trains to Carlisle. During the First World War they were increasingly used on the heavier slower passenger services including troop trains and after the war passenger work predominated over freight turns. At the Grouping their distribution was:

St Margaret's:	9882–9884, 9889–9891
Dundee:	9885, 9886
Berwick:	9887, 9888, 9892
Thornton Jn:	9893

By the Grouping, the Eastfield fleet for the West Highland line had long been displaced by the Reid D34 'Glens'. The dozen engines were then basically working local and semi-fast trains out of Edinburgh to Galashiels and Hawick, Berwick, Perth and Dundee. When Berwick closed in 1924 its engines moved to Tweedmouth, 9882 from St Margaret's being exchanged for 9892. The Thornton Junction single member was augmented by 9885 and 9886 from Dundee and their work included 'The Fife Coast Express' from Edinburgh. G.J. Aston timed a D32 on an afternoon Hexham–Newcastle excursion in June 1930. 9892 had a modest load of six coaches, 147/160 tons, but the run was inhibited by signal checks at Wylam and Elswich and took five seconds under forty minutes for the 20.7 miles. The evening return at 9.14pm was plagued by signal stops from Wylam onwards and the journey took almost fifty minutes. St Margaret's D32s were frequently used as assistant engines, piloting 'Scotts' or Reid Atlantics on the climb to Falahill on the Waverley route or Cockburnspath to Grantshouse on the main line to Berwick. An example of such work is shown below:

Edinburgh Waverley–Galashiels, 24.9.1932

9879 *Abbotsford* (C11 4-4-2) + 9890 (D32) assisting to Galashiels

10 chs, 296 tons

10.03am Edinburgh - Carlisle

Miles	Location	Times	Speed	Gradients
0	Edinburgh Waverley	00.00		1/78 F
6.3	Millerhill	08.55	50	1/80, 1/175 R
12	Gorebridge	17.30	27/31	1/70 R, 1/111 R
18	Falahill Box	29.45	26/29	1/70 R, 1/100 R
22.5	Fountainhall	35.25	60	1/100 F, 1/150 F
26.7	Stow	-	56	1/200 F
<u>33.5</u>	<u>Galashiels</u>	<u>48.30</u>	(pilot engine off)	

Dundee's 9886 departing from Edinburgh Waverley through Princes Street Gardens with the 11.40am Edinburgh to Thornton Junction stopping train, 22 June 1936. (MLS Collection)

In 1938, the Tweedmouth allocation was dispersed, 9882 to Haymarket for local freight work and the other pair to Blaydon for local passenger work in the Border Country. They continued doing this after nationalisation though operating from Hawick depot. 9888 was timed by A.J. Middlemass on the 11.41 Hexham–Newcastle in 1939, keeping good time, maximum speed 58mph, but the load was only four coaches. St Margaret's 9890 was timed in July 1937 by D.S. Barrie on a 5-coach Glasgow–Haymarket train via the former Caledonian route, losing nearly nine minutes to Winchburgh Junction, because of a p-way slowing to 35mph near Dullatur and prolonged signal checks at Greenhill Junction, but recovered a couple by running in the low 60s through Ratho to Saughton Junction. Their final allocation in 1948 was:

The last survivor, 62451, at Galashiels with the 7pm to Selkirk, 7 September 1950. (J.D. Darby/MLS Collection)

St Margaret's:	2443 – 2445 (9882 – 9884), 2450, 62451 (9889, 9890), 2453, 2454 (9892, 9893)
Hawick:	2448 (9887), 2449 (9888)
Thornton Jn:	2446 (9885), 2447 (9886)

The St Margaret's engines worked residential traffic into Edinburgh, local services to Dunbar, banking work at Cockburnspath and occasional football and other excursions. The mass introduction of the Thompson B1 class sealed their fate and the last one, 62451, finished its days working from Dunbar.

62451 at Selkirk with the one-coach 7.55pm to Galashiels, 7 September 1950. (J.D. Darby/MLS Collection)

The NBR Class 'K' (LNER D33)

By 1909, more mixed traffic engines with 6ft coupled wheels were required, but instead of building more of the successful 1906 'Intermediates', Reid decided to build some of the 'Scotts' then under construction (the LNER D29 class) with 6ft instead of 6ft 6in driving wheels. The differences with the 'Intermediates' were small, mainly boiler dimensions and the provision of a larger tender. Although classified 'K' like the 'Intermediates' (LNER D32) they were also confusingly known as 'Intermediates' though it probably mattered little as they shared the same work as the 1906 engines. Twelve engines were constructed at Cowlairs Works at the end of 1909 and beginning of 1910, following on from the construction of the first five 'Scotts'. They were numbered (constructed in this order) 331, 864–867, 894, 332, 333, 382–385. Their dimensions were:

Cylinders (2-inside):	19 x 26in
Coupled wheel diameter:	6ft 0in
Bogie wheel diameter:	3ft 6in
Boiler Pressure:	190lbs psi (later 180lbs psi)

Stephenson motion with piston valves

Heating surface:	1,618.12sq ft
Grate area:	21.13sq ft
Axleload:	19 tons 3 cwt
Weight (Engine):	54 tons 1 cwt
(Tender):	44 tons 12 cwt
(Total):	98 tons 13 cwt
Water capacity:	4,235 gallons
Coal capacity:	7 tons
Tractive effort:	21,953lbs (later 19,945lbs)

Although successful, like the earlier 'Intermediates' further construction was interrupted by Cowlairs Works resuming the building of more 6ft 6in 'Scotts' and then superheating was under serious consideration and further engines with the smaller coupled wheel diameter would be superheated (the 'Glens').

At the Grouping, their NBR livery was replaced by black with red lining, then like the earlier 'Intermediates' changed to the LNER lined green only to revert to the black livery in 1928. They were classified as D33 and their numbers were increased by 9000 becoming 9331–9333, 9382–9385, 9864–9867 and 9894. 9332 was superheated in March 1926, and the rest followed, the last, 9866, not being superheated until April 1936. The Robinson superheater was utilised and the smokebox lengthened by seven inches. Again there was a mixture of engines provided with a Wakefield mechanical lubricator and the Detroit hydrostatic equipment (9332 and 9894). Like other NB 4-4-0s their smokebox wingplates were removed after 1921.

They were allocated 2455–2466 in the LNER renumbering scheme, though not in the order of construction, the first, 331, being grouped with 332 and 333 and following the 864 series. 2456 (ex 9865) and 2465 (ex 9384) were

Below left: **D33 9864,** built in 1909 and superheated in 1928, seen at Stirling, 8 September 1936. (MLS Collection)

Below right: **D33 9866,** built in 1909 and not superheated until 1936, seen here at Perth shed, 9 April 1939. Its Wakefield mechanical lubricator is visible on this side. (MLS Collection)

D33 62457 (former 9866), one of the last three survivors, in BR lined black mixed traffic livery, at Haymarket shed, 21 October 1951. (MLS Collection)

A northbound express from Edinburgh passing Inverkeithing behind a 1909 built 'Intermediate' in the 86X series, c1911. (G.M. Shoults/MLS Collection)

withdrawn in 1947, but the rest were received into British Railways stock and were numbered 62455 – 62466. Withdrawals continued steadily in the early years of nationalisation, the last survivors being 62457 and 62462 withdrawn in 1953 and 62463 (the former 383) withdrawn in September 1953.

Operations

The 'Intermediates' were involved in very similar work to the 1906 6ft 4-4-0s although with the increase in number more depots could receive an allocation. Bathgate, Stirling, Thornton Junction, St Margaret's and Aberdeen were added to Eastfield, although the latter with both classes of 'Intermediate', now used more on the West Highland line, including passenger work, which lasted until enough of the 1913 built 'Glens' were available to displace them.

D33 62459 at Dunfermline station, its home depot, with a stopping train for Edinburgh, 15 September 1950. It has acquired its BR number and smokebox numberplate, but the tender is still inscribed NE as would have been painted during the war years or in the immediate aftermath.
(G.M. Shoults/MLS Collection)

In 1910, the NBR conducted some tests with the 'Scott' and 'Intermediate' classes and another test compared the performance of a 1909 'Intermediate' with a Highland Railway 'Castle' 4-6-0. The tests were between Blair Athol and Dalwhinnie on the Highland and Perth and Kinross on the North British. Apparently, the NB engine was superior and gave Reid confidence to develop the class further.

After the Grouping, with the 'Glens' now numbering thirty-two, the D33s were more widely dispersed. In the mid-1930s the allocations were:

Eastfield:	9331, 9333, 9382
Bathgate:	9332, 9384
Thornton Jn:	9383
Stirling:	9385, 9864, 9894
Perth:	9865, 9866
Hawick:	9867

All worked into Edinburgh from these locations and between Glasgow, Stirling, Dundee and the Fife Coast. The final allocation of those remaining in mid-1950 was:

Dundee:	62457, 62466
Dunfermline:	62459, 62464
Eastfield:	62460, 62462
Stirling:	62461

Both the Dundee D33s moved to Dunfermline within the year, but the rest remained as allocated above until their withdrawal between 1951 and 1953.

The NBR Class 'K' (LNER D34)

These thirty-two 6ft wheeled 4-4-0s were the final development of the North British mixed traffic engine, superheated from the beginning and the mixed traffic equivalent of the 6ft 6in D30/2 'Scotts'. They were known as the 'Superheated Intermediates' but as they were all named after Scottish glens, they became universally known as the 'Glen' class. Just ten were built at Cowlairs initially in 1913, five, numbered 149, 221, 256, 258 and 266 with Schmidt superheaters and the other five, 307 and 405–408, with Robinson superheaters. Five more were constructed in 1917 with Robinson superheaters which became standard for the class, 100, 291, 298, 153 and 241, five more in 1919, 242, 270, 278, 281 and 287 and a final dozen in 1920, numbered 502–505, 34, 35, 490 and 492–496. Their dimensions were:

Cylinders (2-inside):	20 x 26in
Coupled wheel diameter:	6ft 0in
Bogie wheel diameter:	3ft 6in
Boiler pressure:	165lbs psi (later 180lbs psi)

Stephenson motion with 10" piston valves

Heating surface:	1,346.06sq ft (incl superheater 192.92sq ft)
Grate area:	21.13sq ft
Axleload:	19 tons 2 cwt
Weight (Engine):	57 tons 4 cwt
(Tender):	46 tons 13 cwt
(Total):	103 tons 17 cwt

Water capacity: 4,235 gallons
Coal capacity: 7 tons
Tractive effort: 22,100lbs (with 180lbs psi boiler)

The 1913 series were named:

149 *Glenfinnan*
221 *Glen Orchy*
256 *Glen Douglas*
258 *Glen Roy*
266 *Glen Falloch*
307 *Glen Nevis*
405 *Glen Spean*
406 *Glen Croe*
407 *Glen Beasdale*
408 *Glen Sloy*

The 1917 series were named:

100 *Glen Dochart*
291 *Glen Quoich*
298 *Glen Shiel*
153 *Glen Fruin*
241 *Glen Ogle*

The 1919 series were named:

242 *Glen Mamie*
270 *Glen Garry*
278 *Glen Lyon*
281 *Glen Murran*
287 *Glen Gyle* (erroneously painted *Glen Lyon* Nov & Dec 1941)

The final 1920 series were named:

503 *Glen Arklet*
504 *Glen Aladale*
490 *Glen Dessary*
502 *Glen Fintaig*
505 *Glen Cona*
34 *Glen Garvin*
35 *Glem Gloy*
492 *Glen Gau* (corrected to *Glen Gaur* in 1925)

Above: **149** *Glenfinnan*, the prototype 'Glen' as built at Cowlairs in September 1913, allocated to Eastfield and seen at Fort William, c1913. (Locomotive & General/MLS Collection)

Below: **9503** *Glen Arklet*, first of the 1920 built series, at Hawick shed, 7 August 1937. The Wakefield mechanical lubricator is very clear in this photograph. (MLS Collection)

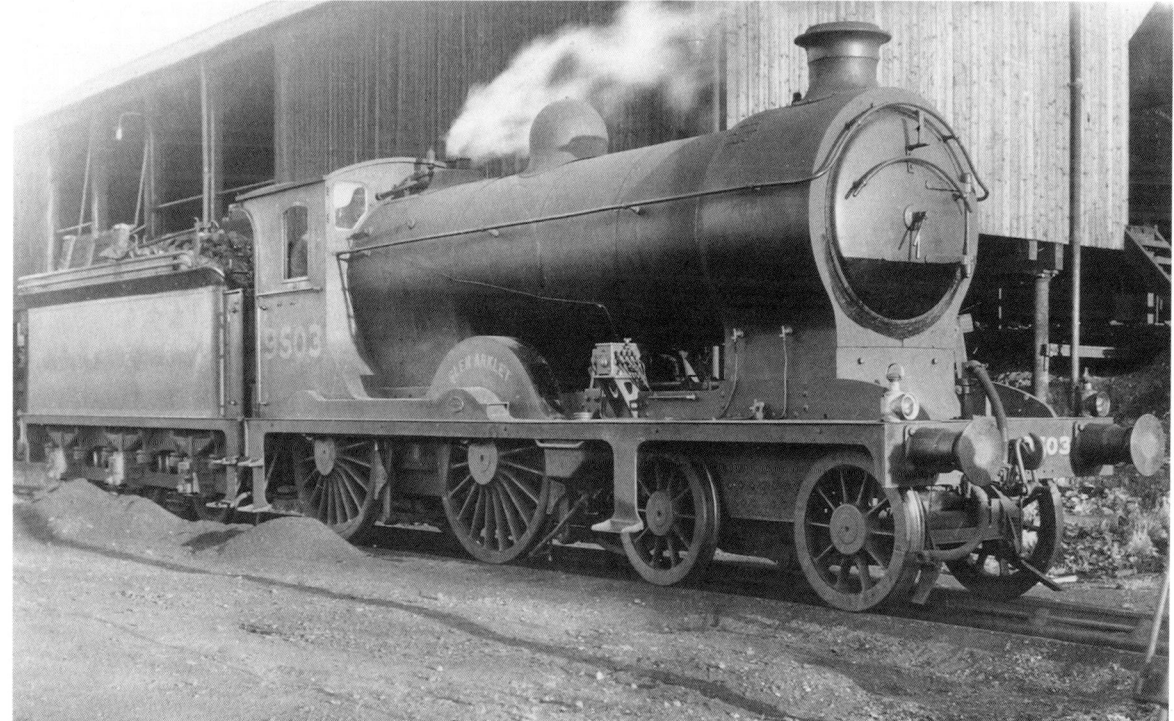

9503 *Glen Arklet,* seen from the other side, with another D34 'Glen' at St Margaret's shed, c1938. (G. Harrop/MLS Collection)

Below left: **9406** *Glen Croe,* after the removal of its Westinghouse air brake pump, at Fort William shed, 12 June 1936. (MLS Collection)

Below right: **2481** *Glen Ogle* at Thornton Junction, 6 April 1947. It was withdrawn in 1949 without receiving its BR number. (MLS Collection)

493 *Glen Luss*
494 *Glen Loy*
495 *Glen Mallie*
496 *Glen Moidart*

The first five 'Glens' had their Schmidt superheaters replaced by the Robinson type and other than reboilering, were subjected to no rebuilding. They were classified as D34 by the LNER and numbered with the additional 9000 when repainted at the Grouping. Their somokebox wingplates were removed between 1922 and 1924. Half of the class had their bufferbeams drilled for carrying small snowploughs and all bar 221 had Wakefield mechanical lubricators. 221 like other individual engines in the other Reid 4-4-0s had the Detroit hydrostatic lubricator. The 1917 – 1920 'Glens' were built with smokebox door handrails and these were added to the 1913 engines subsequently. They were curved in the standard NB way except for 2472 (ex 307) after 1946 which had a straight one. All were dual brake fitted, the Westinghouse air brake being removed between 1935 and 1937.

In 1946, the class was allocated the numbers 2467–2498, but 9287 was withdrawn in February 1946 before carrying its new number (2486) and 2491 (ex 9505) was withdrawn in December 1947, the other thirty being taken into British Railways stock in January 1948 and were renumbered 62467–62498. 2473, 2476 and 2481 were withdrawn in 1949 or 1950 and did not carry their BR allocated number but the rest did and were repainted in the BR mixed traffic lined black livery. These twenty-seven engines then continued in traffic until further withdrawals started in 1958 with

The North British 4-4-0s • 109

62468. Twelve went in 1959, eight in 1960, leaving 62469, 62474, 62479, 62484, 62496 and 62496 as the final survivors, which were withdrawn in 1961. The last two in normal traffic were 62484 *Glen Lyon* and 62495 *Glen Loy*, although 62469 *Glen Douglas*, withdrawn in 1959, had been reserved for preservation, repainted as 256 in NBR livery and was retained for special train working for a few years before being placed in the Glasgow Museum of Transport (see Chapter 7, page 187).

Operations

The prime work of the 'Glens' was the West Highland line, both sections – Glasgow–Fort William, and Fort William–Mallaig. This lasted from their introduction up until nationalisation by which time they were being replaced by the Thompson B1s and then in the 1950s by the former LMS 'Black Fives'. By the completion of the delivery of the 'Glens' in 1920, three-quarters of them were at Glasgow Eastfield. They would be serviced at the Fort William depot or its sub-shed, Mallaig. The maximum load for a 'Glen' on the West Highland was 190 tons, which led to frequent double-heading of the D34s (and the B1s and Black 5s that followed them). There was an attempt to introduce the Gresley K3s to reduce the double-heading, but their axleload was too heavy and a number of the earlier K2s were moved to Scotland in 1925, but even their maximum tonnage of 220 tons was insufficient to rule out much of the double-heading. The design of the K4 2-6-0 that could haul 300 tons came in 1937, too few and too late and the two V4 2-6-2s built in 1941 were much too late, so the D34s soldiered on sharing the work with the K2s and K4s until the B1s arrived.

Above left: An Eastfield D34 62479 *Glen Shiel* in the BR livery with 'lion & wheel' totem on the tender at its home depot, alongside an ex LMS 'Black Five' then beginning to take over the West Highland services from the D34s and LNER B1s, June 1950. (J. Davenport/MLS Collection)

Above right: 62490 *Glen Fintaig* at St Margaret's alongside two B1s, 61356 and 61357, which were among the arrivals in Scotland taking work off the former NB 4-4-0s, 31 August 1952. (MLS Collection)

Left: **62484** *Glen Lyon* in the final BR livery applied to the D34s, at Perth, 6 April 1958. 62484 would be one of the last two survivors withdrawn in November 1961.

9221 *Glen Orchy* leads a second unidentified D34 on the northbound *Northern Belle*, climbing to Beasdale summit with the Glenfinnan viaduct in the background, c1930. (Rail Archive Stephenson/John Scott-Morgan Collection)

Two unidentified 'Glens' near Ardlui with the *Northern Belle* in the 1930s. (Colling Turner/ MLS Collection)

The Eastfield engines also worked the main line to Glasgow, to Stirling and Perth and to the Fife Coast. They were hard working engines, reliable and robust, averaging two years between Works heavy overhauls. After the Grouping at least half of the class remained at Eastfield, the others split between St Margaret's and Thornton Junction. The actual locomotive allocation in 1923 was:

Eastfield:	9100, 9153, 9221, 9241, 9242, 9256, 9258, 9281, 9298, 9307, 9405–9408, 9490, 9493–9406
St Margaret's:	9266, 9270, 9278, 9287, 9492, 9502–9504
Thornton Jn:	9034, 9035, 9149, 9291
Dundee:	9505 (later moved to Thornton Jn)

During the 1930s, a few moves were made – 9221 and 9407 from Eastfield to Thornton Junction in return for 9035 to Eastfield and 9493 was based at Haymarket for a while working the Edinburgh–Crail–Dundee service. The Thornton engines worked to Glasgow and Edinburgh and the Fife Coast. The St Margaret's engines tended to work to the south or west of Edinburgh, including the main lines to Berwick and Carlisle and intermediate stopping services. The 'Glens' were used frequently on relief trains, excursions and specials.

On the West Highland line two 'Glens' worked *'The Northern Belle'* splitting at Craigendoran, but such was the load that the portions were worked forward by pairs of 'Glens'.

O.S. Nock logged a number of runs on the West Highland line in 1934 and a couple of these are shown in the tables below:

		Fort William–Craigendoran, 1934 5.12pm Fort William – King's Cross Sleeper service					
		9494 *Glen Loy*		9035 *Glen Gloy*			
		182/190 tons		182/190 tons (max load)			
Miles	Location	Times	Speed	Times	Speed		Gradients
0	Fort William	00.00		00.00			
3.4	MP 96 ½	07.15	23½	07.14	27½		1/59 R
8.4	MP 91 ½	15.07	47½	14.29	53		1/80, 1/165 R, 1/100 F
9.5	Spean Bridge	16.30		15.37			
0		00.00		00.00			
1.4	MP 89	03.35		03.49			1/188 R
3	Roy Bridge	-	40	-	39½		1/84 R
4.4	MP 86	09.10	26	09.06	26		1/64 R
6.4	MP 84	12.45	39	12.45	37½		1/126 R
8.4	MP 82	16.37	26½	17.13	22		1/59 R
8.7	Tulloch	17.20		17.48			
0		00.00		00.00			
3	MP 78 ¾	07.30	37½	07.35	41½		1/59 R, 1/444 R
6.7	MP 75	15.17	25	15.11	22½		1/67 R
8.7	MP 73	19.40	32½	19.55	32		1/67 R
10	Corrour	22.25	25½	22.41/23.35			1/57 R, 1/140 F
14.7	MP 67	-	54	-	62½		1/79 F
17.3	Rannoch	32.45	1¼ E	33.18		¾ E	
0		00.00		00.00			
6.7	Gortan	10.15	53	10.35			1/148 R, L
15.4	Bridge of Orchy	22.00	60/25*	21.33	58½/30*		1/56 F, 1/74 F
19.2	MP 45	29.25	31½	28.15	37		1/132 R, 1/254 R
21.2	MP 43 (summit)	34.00	25	32.48	23		1/57 R
23.1	Tyndrum	36.25	52	35.15	50		1/60 F
28.1	Crianlarich	43.30	3½ E	42.25		4½ E	
0		00.00		00.00			
16.7	Arrochar	30.55		28.57			
0		00.00		00.00			
4.4	Glen Douglas	11.05	23	11.46	20½		1/57 R
10.6	Garelochhead	20.45	56	22.48	easy/braking		1/80, 1/54 F
19.6	Craigendoran	37.25		37.58			

Both trains were then amalgamated with other sleeper portions and proceeded to Glasgow Queen Street. Nock also timed a northbound heavy train of 365 tons double-headed by a couple of D34s:

Helensburgh–Fort William, 1934
9035 *Glen Gloy* + 9221 *Glen Orchy*
341/365 tons

Miles	Location	Times	Speed		Gradients
0	Upper Helensburgh	00.00		T	
1.9	Rhu	-			1/67, 1/88 R
6.9	Garelochhead	12.05			L, 1/92 F
0		00.00			
1.3	Whistlefield	05.05	23	(thick mist)	1/58 R
6.2	Glen Douglas	14.58	37½		1/80 R, 1/122 R
10.6	Arrochar	21.51			
0		00.00			
8	Ardlui	12.30	35		
15.4	MP 35 (summit)	27.35	26		1/60 R
16.7	Crianlarich	30.35		3 E	
0		00.00			
5	Tyndrum	11.17	26		1/60 R
6.7	MP 43 (summit)	15.05	28½		1/83, 1/63 R
12.5	Bridge of Orchy	23.06	20*		1/55 F
15.7	MP 52	26.55	42½		1/240 R
21.2	Gortan	38.53	27		1/66 R
28.1	Rannoch	48.00			
0		00.00			
2	MP 66 ¼	-	21		1/53 R
4.7	MP 69	-	46		1/83 F, 1/191 R
7.3	Corrour	14.17	sigs sl (xing)		
0		00.00			
8.5	Fersit	14.12			1/67 F
0		00.00			
1.2	Tulloch	03.25			1/67 F
0		00.00			
5.7	Roy Bridge	09.57			1/59, 1/126, 1/64 F
0		00.00			
3	Spean Bridge	06.00			1/188 F
0		00.00			
8.5	Mallaig Junction	11.35	59		1/80, 1/59 F
9.5	Fort William	14.05		T	

The 'Glens' continued on the same type of work apart from *'The Northern Belle'* and excursion work during the Second World War and one turn that saw a 'Glen' stabled at Perth with a long turnover often found it utilised on the Highland line as far as Blair Athol. When the Thompson B1s began to be allocated to Scotland, a number of allocation changes took place, and in 1953 eight of the Eastfield engines were sent further north to replace withdrawn Great North of Scotland 4-4-0s. The allocations in 1950 and 1953 were:

1950
Eastfield: 62469, 62470, 62472, 62474, 62477, 62479, 62480, 62482, 62489, 62493, 62496–62498
St Margaret's: 62471, 62483, 62484, 62487, 62488, 62490, 62494
Thornton Jn: 62467, 62468, 62475, 62478, 62492
Dundee: 62485
Bathgate: 62495

December 1953
Eastfield: 62472, 62474, 62477, 62496, 62498
St Margaret's: 62471, 62483, 62487, 62488, 62490, 62494
Perth: 62470, 62484
Thornton Jn: 62467, 62468, 62475, 62478, 62492
Dundee: 62485
Kittybrewster: 62469, 62479, 62480, 62482, 62489, 62493, 62497, 62498

Above: 9242 *Glen Mamie* and K2 4698 *Loch Rannoch* on the turntable at Mallaig, 31 July 1938. (MLS Collection)

Below: 62474 *Glen Croe* leaving Perth with a stopping train for Glasgow, 27 May 1953. (MLS Collection)

62489 *Glen Dessary* transferred to Kittybrewster in February 1953, seen here after taking water at Aberdeen Ferryhill shed, 1957.
(S. Bryant/MLS Collection)

62469 moved to Keith in February 1956 before its withdrawal and preservation in 1959 and 62483 and 62494 were transferred from St Margaret's to Hawick in February 1958 and 62488 followed in November 1959, otherwise their allocations remained static until withdrawal. At the end of 1960, just 62474, 62479, 62484, 62488, 62495 and 62496 remained. The Kittybrewster engines were employed mainly on goods traffic between Aberdeen, Keith and Inverness and on the Deeside lines. Two D34s, 62471 and 62496, spent a week on the West Highland line in May 1959 being filmed for the TV programme, *Railway Roundabout,* and recorded by Peter Handford for the Transacord railway sound LPs. The last in regular traffic were 62484 and 62496, withdrawn in November 1961, although 62469, as NB 256, remained for special working after its restoration until its official withdrawal in December 1962.

Eastfield's 62472 *Glen Nevis* with the empty stock of a half day excursion from Glasgow at Balloch Pier, Loch Lomond, with V1 2-6-2T 67633, 6 June 1953.
(MLS Collection)

62496 Glen Loy working on the West Highland line in May 1959 during a week when being filmed with 62471 on the TV *Railway Roundabout* programme. (MLS Collection)

62484 Glen Lyon on arrival at Carlisle with the 12.25pm stopping train from Hawick, alongside a 'Black 5' and 'rebuilt Patriot', 1 July 1961. (MLS Collection)

The NBR Class 'N' (LNER D35)

Twenty-four 4-4-0s with 5ft 7in coupled wheels were built by Matthew Holmes between 1894 and 1896 specifically for the West Highland line. 693–701, 55, 394, 395 were constructed in 1894 and 341–346, 227, 231, 232 and 702–704 were built in 1896. They were all constructed at Cowlairs. Their dimensions were:

Cylinders (2-insde):	18 x 24in
Coupled wheel diameter:	5ft 7in
Bogie wheel diameter:	3ft 6in
Boiler pressure:	150lbs psi

Stephenson motion with slide valves

Heating surface:	1,235.13sq ft
Grate area:	17sq ft
Axleload:	14½ tons
Weight (Engine):	43 tons 6 cwt
(Tender):	32 tons
(Total):	75 tons 6 cwt
Water capacity:	2,500 gallons
Coal capacity:	6 tons
Tractive effort:	14,798lbs

They were known as 'The West Highland Bogies', but performance fell short of expectations and 695 was rebuilt in 1919 as Class 'L' (LNER D36 – see next section). However, no further rebuilding of the other twenty-three took place and sixteen were withdrawn between 1919 and 1922, leaving seven which were transferred to the duplicate list as 1434 (55), 1439 (342), 1442 (345), 1448 (696), 1449 (697), 1452 (701) and 1453 (704). These were allocated class D35 and were renumbered at the Grouping 10434, 10439, 10442, 10448, 10449, 10452 and 10453. Four of these were withdrawn in 1923 without receiving these revised numbers, 10453 similarly in January 1924. The last two, 10439 and 10448 were allocated LNER numbers 9997 and 9998, but again, these numbers were never carried, both locomotives being withdrawn in the autumn of 1924.

Operations

The engines were allocated to Cowlairs and Fort William for working on the West Highland line as intended. The Mallaig extension was opened in 1901 and one, usually 344, was based there long after the rest had been replaced by the earlier Drummond 'Abbotsford' class (D27) & D28). The Cowlairs engines were also used on local services from Glasgow along Clydeside and commuter traffic to Helensburgh and Balloch. They moved to Eastfield when the new depot replaced Cowlairs and several moved to Parkhead for local services and stopping trains to Edinburgh via Bathgate. A few moved to St Margaret's for passenger services to Dundee and Perth.

697 and 698 worked in the Galashiels/Peebles area and 231 and 700 also worked from Galashiels on banking and assistance duties. 232 was stationed at Hawick for working to Edinburgh. During the First World War, 55 was working the Fort Augustus branch replaced by 693 in 1920. During the war, the St Margaret's engines were used for piloting heavy war traffic trains and four were sent to Dunfermline (55, 343, 346 and 703) for working at Inverkeithing. 231 finished at Berwick and 697 at Carlisle.

231, then based at St Margaret's, leaving Aberdour with a stopping train for Dundee, c1911. (G.M. Shoults/MLS Collection)

232 passing the King Alexander III monument at Kinghorn, Burntisland, with a stopping train, c1911. (G.M. Shoults/MLS Collection)

After the Grouping the remaining engines were working from:

1434:	Dunfermline (as Inverkeithing pilot)
1439:	Haymarket
1442, 1448, 1449, 1453:	Eastfield for local passenger services
1452:	Bathgate for stopping trains to Edinburgh.

Their poor reputation, particularly on the West Highland line, was attributed to their low adhesion, causing much slipping on the steep banks of the route.

The NBR Class 'L' (LNER D36)

In February 1919 Reid undertook the drastic rebuilding of 695, one of the 'West Highland Bogies' which had such a poor reputation. It was virtually a new engine – new frames, cylinders, superheated boiler and a Reid cab. Perhaps it was intended to rebuild the rest, but the cost must have been nearly as high as the construction of a new engine, and the success of the rebuilding was clearly insufficient to warrant more. In 1919 and 1920, Reid got his Board to authorise seventeen more 'Glens' instead. The rebuild's dimensions were:

Cylinders (2-inside):	19 x 26in
Coupled wheel diameter:	5ft 7in
Bogie wheel diameter:	3ft 6in
Boiler pressure:	165lbs psi
Stephenson motion with 10in piston valves	
Heating surface:	1,282.3sq ft (incl 220sq ft superheating)
Grate area:	19.5sq ft
Axleload:	17 tons 9 cwt
Weight (Engine):	49 tons 11 cwt
(Tender):	33 tons 10 cwt
(Total):	83 tons 1 cwt
Water capacity:	2,500 gallons
Coal capacity:	6 tons
Tractive effort:	19,648lbs

The new boiler was fitted with a Robinson superheater with extended smokebox and lasted to 1936, when it was replaced by a D31 saturated steam boiler, which it retained until its withdrawal in May 1943. Its smokebox wingplates were removed in 1923. It had a Wakefield mechanical lubricator and in 1925, despite the recent

9695 after its superheated boiler had been replaced by a D31 saturated boiler in 1936, seen here c1937. Note that it still has the Westinghouse air brake pump at this stage.
(MLS Collection)

rebuilding, its frame was rebuilt to a stronger one, with a higher profile over the running plate. It was fitted with a dual brake system on rebuilding, though the Westinghouse air brake was removed in the mid-1930s.

Operation

It remained at Parkhead, its depot before the rebuilding, working from Bridgeton Cross to Helensburgh and Milngavie. In 1931, it was transferred to Eastfield, where in the summers of the 1930s it piloted excursions to Oban. In winter it worked passenger services from Queen Street to Kinross. During the Second World War, it was noted on lowly duties radiating from both Glasgow and Edinburgh. Despite its solitary state, it amassed a considerable mileage – 1,695,688, equivalent of the average mileages of the contemporary D31s and more than the later 'Glens'. After its withdrawal it was used for several months as a disabled engine in ARP exercises.

Chapter 4
THE GREAT NORTH OF SCOTLAND 4-4-0s

The chronological order of the construction of the Great North of Scotland 4-4-0s was not followed by the LNER in classifying the different groups of engines, and whilst I deal with the classes in the LNER order for easy reference, for clarity I list below the classes in the order in which they were designed, built and rebuilt. The logic of the LNER classification is not apparent.

The Great North of Scotland 'Q' class (LNER D38)

Manson designed and built a number of 4-4-0s in the 1880s, three small classes with coupled wheel diameter of 5ft 6in to 6ft, then, because of increasing train loads, larger 4-4-0s in 1888-1890, with 6ft and finally 6ft 6½in driving wheels, the latter as class 'Q' and later classified by the LNER D38, the earlier engines becoming classes D42-D44, D46 and D48. These 1890 larger wheel engines were just three in number, a repeat of the 1888 class 'O' with the 6in larger diameter coupled wheels and unique for the railway, with eight-wheel larger tenders. They were numbered 75 – 77 and, with the Class 'Os', built by Robert Stephenson & Co at a cost of £2,525 each (£25 more than the 'Os' - extra for the larger wheels). Their dimensions were:

LNER class	GNoS class	Wheel Diameter	Built	Designer	Rebuilt
The D47/2	K	5' 6 ½"	1866	Cowan	1889-1891
The D47/1	L	5' 6 ½"	1876	Cowan	1890-1899
The D39	C	6' 1"	1878	Cowan	
The D45	M	5' 7"	1878	Cowan	1896-1903
The D44	A	6' 0"	1884	Manson	1905-1912
The D48	G	5' 6"	1885	Manson	1905-1911
The D46	N	5' 7"	1887	Manson	
The D42	O	6' 0 ½"	1888	Manson	1915-1920
The D43	P	6' 0 ½"	1890	Manson	1916
The D38	Q	6' 6 ½"	1890	Manson	
The D41	S	6' 1"	1893	Johnson	
The D41	T	6' 1"	1895-1898	Pickersgill	
The D40	V	6' 1"	1899-1915	Pickersgill	
The D40	F	6' 1"	1920	Heywood	

Cylinders
 (2-inside): 18 x 26in
Coupled wheel
 diameter: 6ft 6½in
Bogie wheel
 diameter: 3ft 9½in
Boiler pressure: 165lbs psi

Stephenson motion with slide valves
Heating surface: 1,165sq ft
Grate area: 18.2sq ft
Axleload: 16 ½ tons
Weight (Engine): 46 tons
 (Tender): 37 tons 8 cwt
 (Total): 83 tons 8 cwt
Water capacity: 3,000 gallons
Coal capacity: 5 tons
Tractive effort: 15,050lbs

One of the three class 'Q', believed to be 75, as built in 1890 with the unique eight-wheel tender, seen here c1900. (Photomatic/John Scott-Morgan Collection)

Below left: 77s as rebuilt with superheated boiler and extended cab in 1913, retaining the GNoS smokebox door but with six-wheel tender, seen after the Grouping before more radical renumbering, c1924. (Real Photographs/MLS Collection)

Below right: 6877 in its latter days in LNER Black livery at Keith shed, 10 August 1937. It was withdrawn just a month later. (MLS Collection)

77 was fitted with a new superheated boiler (Schmidt type) in 1913 and 75 was equipped with a Robinson superheater boiler in 1917. 76 had been given a replacement saturated steam boiler in 1914. The superheated boilers had 1,000sq ft of heating surface of which the superheater contributed 140sq ft. The boiler pressure was reduced slightly to 160lbs psi, but the engine weight was increased by exactly a ton and the axleload by 3 cwt. Because of the reduction in boiler pressure, the tractive effort declined slightly to 14,594lbs. After the grouping they were renumbered 6875–6877. Their smokebox doors were replaced by North British style ones as it was difficult to keep the GNoS ones steam tight. The cab

roofs were extended on 76 and 77 during the rebuilding and a pillar handrail added. Their livery was changed to the LNER lined green after the Grouping, but reverted like all the pre-grouping 4-4-0s to lined black after 1928.

The eight-wheeled tender had a leading bogie and rigid framed pair of axles at the rear. Their increased coal and water capacity compared with other GNoS 4-4-0s enabled through working between Aberdeen and Elgin without need for water or coal replenishment. 75 and 77 had lost their eight-wheel tenders before the Grouping and 76 exchanged its for a six-wheel tender in early LNER days. The eight-wheel tenders were scrapped.

The unsuperheated 6876 was withdrawn in 1931, but the other two survived to the late 1930s, 6877 going in September 1937 and 6875 in January 1938.

Operation

The three locomotives were based at Kittybrewster and worked the express services for which they'd been designed, namely, expresses from Aberdeen to Elgin. After the Grouping 6875 remained at Kittybrewster, but 6876 moved to Keith and 6877 to Elgin, still working trains between those locations and Aberdeen. In the 1930s both returned to Kittybrewster, 6876 being withdrawn, but the two superheated engines remained active on the Banchory and Ballater branches and occasional specials which included excursions to Boat of Garten on which 6875 in particular was reputed to be a favourite.

The Great North of Scotland 'C' class (LNER D39)

We move back in time to the 1870s and the era of William Cowan. The Directors ordered twelve 4-4-0s designed by Cowan from Neilson & Co. in 1877, nine being fitted with 5ft 7in coupled wheels (see later Class 'M', D45), the other three with 6ft 1in coupled wheels. These latter three were numbered 1 – 3 and classified 'C'. They were, unlike most 4-4-0 designs of the nineteenth century, outside cylinder engines. Their dimensions were:

Cylinders (2-outside):	17½ x 26in
Coupled wheel diameter:	6ft 1in
Bogie wheel diameter:	3ft 0½in
Boiler pressure:	140lbs psi
Stephenson motion with slide valves	
Heating surface:	1,107.4sq ft
Grate area:	14sq ft
Axleload:	14 tons 3 cwt
Weight (Engine):	41 tons 5 cwt
(Tender):	29 tons
(Total):	70 tons 5 cwt
Water capacity:	1,950 gallons
Coal capacity:	4 tons

Pickersgill rebuilt these engines with boilers that were also used in the rebuilding of the classes 'L' and 'M', with 1,118.5sq ft of heating surface, an enlarged grate area of 15.64sq ft and boiler pressure of 150lbs, raising the tractive effort to 13,907lbs. Nos. 1 and 3 were rebuilt in 1897 and 1898 respectively and 2 later in 1904.

After the Grouping, No.1 was withdrawn in 1925 before renumbering, but the other two

Cowan outside cylinder class 'C' No.3 built by Neilson & Co. in 1879, seen here in original condition, c1905. (Photomatic/MLS Collection)

Elgin's 6803 on a local branch train, c1925. (Robert Fysh/MLS Collection)

were renumbered 6802 and 6803 and were classified as D39. 6802 was withdrawn in 1926 and 6803 in 1927.

Operation

These three locomotives were used on the Aberdeen–Elgin expresses until the introduction of Manson 4-4-0s ten years later. They still worked expresses occasionally thereafter, but eventually worked the Deeside lines. By the time of the Grouping, they were relegated to local and branch line work. No.1 was the Kittybrewster spare engine or Inverurie pilot before its withdrawal. 6802 at Keith and 6803 at Elgin worked local branch trains.

Great North of Scotland Classes 'V' and 'F' (LNER D40)

In 1898, Pickersgill sought authority to contract for twelve new 4-4-0s to cope with increasing traffic, and the Neilson, Reid company offered to build more engines similar to the ones recently delivered (the 'T'class, LNER D41). The new engines varied by having side-window cabs, but after only five had been delivered, the order was terminated as the railway company was having financial difficulties and the remaining locomotives of the order were purchased by the South Eastern & Chatham Railway. The locomotives delivered in 1899 and were numbered 113–117, although 116 and 117 were renumbered 25 and 26 in January 1900. They were classified 'V'. Their dimensions were:

Cylinders (2-inside):	18 x 26in
Coupled wheel diameter:	6ft 1in
Bogie wheel diameter:	3ft 9½in
Boiler pressure:	165lbs psi
Stephenson motion with slide valves	
Heating surface:	1,172.5sq ft
Grate area:	18.24sq ft
Axleload:	16 tons 18 cwt
Weight (Engine):	46 tons 7 cwt
(Tender):	37tons 8 cwt
(Total):	83 tons 15 cwt
Water capacity:	3,000 gallons
Coal capacity:	5 tons
Tractive effort;	16,184lbs

In 1903, Pickersgill sought authority for a further ten locomotives, but only four were ordered in 1906 and these were not built at Inverurie Works until 1909/10. They were numbered 27, 31, 36 and 116, the latter being renumbered 29 in July 1909. The renumbering of this and the earlier engines arose due to accounting decisions, the 113 series being charged to the capital account and the lower numbers replacing withdrawn engines to the revenue account. Pickersgill tried again in March 1911, seeking a further eight, but once more the directors were cautious and cut the authorisation to four, which were built at Inverurie between 1913 and 1915. They were numbered 28 and 33–35.

After withdrawal of older classes at the end of the First World War, a further replenishment of stock was required. Pickersgill's successor turned again to his predecessor's design but the eight locomotives, six built by the North British Locomotive Company and two at Inverurie Works, were superheated. Because of this they were reclassified as 'F'. The NB engines were numbered 47–50,

52 and 54 and were delivered in 1920. The Inverurie built engines, constructed in 1921 were 45 and 46. The superheated boiler had 1,000sq ft of heating surface including 140sq ft provided by the superheater, increasing the engine weight by over two tons to 48 tons 13 cwt. Other dimensions were identical to those of the 'V' class.

After the Grouping, the 1899 engines were renumbered 6825, 6826, 6913–6915, the 1909/15 engines 6827–6829, 6831 and 6833–6836 and the 1920/21 engines 6845–6850, 6852 and 6854. The LNER classified both the 'V' and 'F' classes as D40, without the subdivision used on other pre-Grouping classes where separate identities were given to saturated and superheated engines. Little alteration was made to any of these engines during their career other than the needed reboilering, no decisions being made to superheat any of the earlier saturated steam locomotives. In fact, two superheated engines, 6847 for three years from 1928 and 6848 for a short time in 1946/7, bore a saturated boiler, but reverted to a superheated one at the next Works visit. The 'V' class were fitted with Westinghouse air brakes, with three – 26, 113 and 115 – being supplied with the vacuum brake also for working Highland Railway stock between Aberdeen and Inverness. Tender cabs for tender-first working on branch lines were carried by 6825, 6828, 6829 and 6831 in LNER days and by five of the class in the BR era.

There was no tradition of naming locomotives on the Great North of Scotland Railway, but names

Class 'V' No.25 built by Neilson & Co. to the design of Pickersgill in 1899 (initially numbered 116 but renumbered in January 1900), seen c1900. It became BR 62260 and was not withdrawn until 1953. (Real Photographs/MLS Collection)

Superheated Class 'F' No.54, as built by the North British Loco Company in October 1920. (Railway photographs/MLS Collection)

Superheated Class 'F' No.50 *Hatton Castle* built by the North British Company in 1920. It bears the original GNoS smokebox door, Westinghouse air pump and brass nameplates. (Real Photographs/MLS Collection)

Hatton Castle again, class 'F', now reclassified as a D40 and renumbered 6850 in LNER lined green livery at Kittybrewster shed in September 1930, before repainting black as decreed in 1928. Its GNoS smokebox door has been replaced by a NB one. (J.A.G. Coltas/MLS Collection)

Below left: **D40 6845** *George Davidson* in LNER black livery at Kittybrewster in May 1936. It was built in 1921, became BR 62274 and was withdrawn in 1955. (MLS Collection)

Below right: **Former class 'V'** 6833, built in 1913, at Kittybrewster painted plain wartime black, 5 August 1945. (MLS Collection)

were selected for the Heywood superheated engines and they were:

47 *Sir David Stewart*
48 *Andrew Bain*
49 *Gordon Highlander*
50 *Hatton Castle*
52 *Glen Grant*
54 *Southesk*
45 *George Davidson*
46 *Benachie*

The names were carried on brass plates with raised lettering on the large splasher over the front coupled axle.

In the LNER 1946 scheme, the D40s were renumbered in the series 2260–2280, broadly though not completely in their construction order (see Appendix page 225

for details) and at nationalisation they became 62260–62279, with gaps as 2263, 2266 and 2280 were withdrawn in 1946 and 1947. 62273 kept its brass nameplate until withdrawal but the rest lost theirs in 1954, retaining the names painted on the splashers in small Gill sans lettering. Further withdrawals began in 1953, the class being down to five by 1956, with only 62277 (the former 49) lasting until 1958. It was then preserved and repainted in GNoS pre1920 livery complete with the brass nameplates and retained in running condition until taken into the Glasgow Transport Museum (see Chapter 7).

Former class 'V' No.25, now BR 62260 in full BR lined black mixed traffic livery, at Kittybrewster, 25 May 1953, just three months before withdrawal. Note the tender cab for protection when running tender-first. (MLS Collection)

Class 'V' prototype, 113 of 1899, renumbered 6913 in 1924, 2262 in 1946 and 62262 in 1949, at Keith shed, 26 May 1953. It appears to have some collision damage to the front bufferbeam and running plate. (MLS Collection)

Class 'F' 62279 *Glen Grant* (ex- 52 of 1920) equipped with miniature snowplough at Maud station, 2 June 1953.
(A.C. Gilbert/MLS Collection)

The brass nameplate of 62273 *George Davidson* which survived the loss of the nameplates on the other D40s, keeping its until withdrawal in January1955, seen 26 May 1953.
(MLS Collection)

Operation
Both 'V' and 'F' locomotives undertook most of the express working of the system between Aberdeen and Keith and Elgin. From 1908, their work also included through working to Inverness over the Highland Company's metals. This increased during the First World War with greater military and naval traffic in North Scotland. The express working continued after the Grouping, though from 1925 there was some infiltration of the NB D31 class. Then, in 1931, a number of ex Great Eastern B12 4-6-0s were drafted to the area and took over

most of the heavier express traffic. Most of the D40s were shedded at Kittybrewster with a few at Keith and Elgin.

Records of performance are sparse apart from logs of the preserved engine in the late 1950s. I could find no logs of any express running but one run on a mainline stopping service was recorded by Gerald Aston in 1938:

Elgin–Huntley, 25.8.1938
6914, 6chs 122 tons
9.35am Elgin–Huntley
(Single line, Elgin to Keith)

Miles	Location	Times	Speed		Gradients
0	Elgin	00.00			
1.5	MP 79	-	33/29½		1/165 R, 1/68 R
3.1	Longhorn	07.22		1¼ L	
0		00.00			
2.3	Coleburn Siding	08.19/08.21 sigs		2¼ L	1/50 R, 1/74 R L
3.6	Birchfield	13.48	46		L
5.6	MP 72	-	62½		1/50 F
6.7	Rothes	17.45		2 L	
0		00.00		2¼ L	
1.7	MP 69	-	42		1/80 R
2.1	Craigellachie	05.27		1¾ L	
0		00.00		1½ L	
1.9	MP 65	-	26/31½		1/80, 1/78 R
3.8	Dufftown	09.07		1¼ L	
0		00.00			
1.9	MP 63	-	29		1/60 R
3.9	MP 61	-	54		L, 1/70 F
4.4	Drummuir	08.06		1½ L	
0		00.00			
3	Auchindachy	04.45	46/57		1/146 F
5.7	Keith Town	08.18		1¼ L	
0		00.00			
0.5	Keith	01.43		1 L	
0		00.00			
3.3	MP 51	-	53		L
4.3	Grange	07.02		1 L	
0		00.00		¾ L	
0.8	Cairnie Junction	02.21	44		1/200 F
3.8	Rothiemay	-	52		L, 1/160 R
7.4	Huntley	11.14		T	

6850 *Hatton Castle*
and 6846 *Benachie* head the royal train at Ballater, August 1928.
(John Scott-Morgan Collection)

As well as working the lighter and stopping services on the main line, they worked the subsidiary lines to Buchan, Ballater, Lossiemouth and generally in the Speyside and Deeside. They worked royal trains on the Ballater line, usually double-headed. Frequent excursions in the 1930s from Aberdeen to Boat of Garten were run throughout by Kittybrewster D40s. These weekly half-day excursions were scheduled to run the 67.8 miles from Aberdeen to Craigellachie in 88 minutes, an average speed of 46.2mph, the fastest on the line. Later a Dufftown stop was added, scheduled 81 minutes for the 64 miles, average speed 47.4mph. The load varied between seven and nine coaches, 115–140 tons. The outline times of such an excursion were noted with 'D40' 6825 on 140 tons:

Aberdeen–Craigellachie, c1934

6825, 140 tons

Aberdeen–Boat of Garten excursion

Miles	Location	Times	Speed	Gradients
0	Aberdeen	00.00		
13.4	Kintore	24.00		Rising gradients to Dyce (6 m), then undulating
31.8	Kennethmont	45.04	54 (ave)	Rising with stretches of 1/100 and 1/200
53.8	Keith Town	66.14	62.5 (ave)	Falling with stretches of 1/100 and 1/200, then L
59.6	Drummuir	74.04	44½	Rising, final stretch 1/70
67.8	Craigellachie	85.48	2¼ E	

6850 *Hatton Castle* on the turntable at Kittybrewster shed, 5 August 1945. (MLS Collection)

Another unidentified 'D40' made the 67.8 mile run in 79 minutes 32 seconds (average 51.1mph) but with 112 tons only. 6825 (again) with 105 tons in the reverse direction ran the 7.4 uphill miles from Cairnie Junction to the Huntly stop in 9 minutes 32 seconds and the 8.8 miles to Kennethmont, including 1½ miles climbing at 1 in 100 in 11 minutes and 6 seconds. The 40.8 miles from Huntly to Aberdeen excluding the 7 minutes 5 seconds standing at four intermediate stops took 53½ minutes net. Very occasionally a D40 would venture south of Aberdeen piloting a heavy East Coast express as far as Dundee, usually tucked inside a Reid Atlantic.

After nationalisation, with the Thompson B1s now supplementing the B12s, the remaining D40s were limited to branch line working, some being sub-shedded at locations such as Banff and Macduff. The last workings, especially with the surviving *Gordon Highlander,* were to Boat of Garten on Speyside.

62273 *George Davidson* with a local stopping train at Inverurie, June 1949. (MLS Collection)

62275 *Sir David Stewart* entering Torphin with the 10.20am Ballater – Aberdeen, 18 July 1949. (J.D. Darby/MLS Collection)

62276 *Andrew Bain* entering Macduff with a local train from Inverurie, c1949. (MLS Collection)

The Great North of Scotland 4-4-0s • 131

62264, the former 115 of 1899, on a down freight at Craigellachie, 20 September 1949. (H.D. Bowtell/MLS Collection)

Below left: **62276** *Andrew Bain* on the small shed at Ballater, 18 July 1949. (MLS Collection)

Below right: **62267, the** former 29 of 1909, at Elgin with the Lossiemouth branch train, 26 May 1953. (MLS Collection)

The Great North of Scotland Classes 'S' & 'T' (LNER D41)

James Johnson designed a set of 4-4-0s in 1893 that were built by Neilson & Co., the first six of which were delivered that year and were classified 'S', numbered 78–83. They were the 4-4-0s that Pickersgill later developed as the LNER class D40. Their key dimensions were:

Cylinders (2-inside):	18 x 26in
Coupled wheel diameter:	6ft 1in
Bogie wheel diameter:	3ft 9½in
Boiler pressure:	165lbs psi

Stephenson motion with slide valves

Heating surface:	1,172.5sq ft
Grate area:	18.24sq ft
Axleload:	15 tons 17 cwt
Weight (Engine):	45 tons
(Tender):	37 tons 8 cwt
(Total):	82 tons 8 cwt
Water capacity:	3,000 gallons
Coal capacity:	5 tons
Tractive effort:	16,184lbs

Johnson's successor, Pickersgill, ordered and Neilson's delivered a further fourteen in 1895 and 1896, numbered 93–100 and 19–24, the latter from the revenue account theoretically replacing withdrawn locomotives. Although differences from the earlier machines were minor – such as the type of safety valve, they were classified 'T', as were a further dozen delivered by Neilson's again in 1897 and 1898. These were numbered 101–112, in the capital account. The directors had tendered and Sharp, Stewart had made the most competitive bid, but Neilson's promised the earliest delivery and got the contract.

There was no subsequent rebuilding of the class, other than routine reboilering and the whole class of thirty-two locomotives passed to the LNER at the Grouping and were all, both 'S' and 'T', classified as D41. The six former 'S' engines were renumbered 6878–6883, and the 'Ts', 6819–6824 and 6893–6912, though not in the order of building (see Appendix page 226 for details).

All the locomotives had their GNoS smokebox doors replaced by the NB type and five were fitted with extended smokeboxes between 1928 and 1931 as an experiment (unsuccessful) to reduce coal consumption. The five were 6821, 6898, 6902, 6905 and 6909. Only 6821 retained this until its early withdrawal in 1946. However, 6902 (later 62246) was refitted, retaining this boiler until its withdrawal in 1951. After the

Class 'T' 93 built in December 1895, seen here c1910. (Photomatic/MLS Collection)

The Great North of Scotland 4-4-0s • 133

Class 'S' 79 built in December 1893, with the original GNoS smokebox door, seen c1920. (A.G. Ellis/MLS Collection)

Class D41 6819, built in February 1896, at Elgin, 11 August 1937. It has the NB smokebox door and is in post 1928 LNER black livery, but otherwise unaltered from its construction. (N. Fields/MLS Collection)

One of the original 'S' class, built in 1893, 6883, on the turntable at Kittybrewster, 10 August 1937. It bears obvious signs of burning the smokebox door. (MLS Collection)

6905, built in 1897, in postwar rundown condition, at Keith, 4 August 1945. (N. Fields/MLS Collection)

Grouping all the D41s were dual fitted with Westinghouse and vacuum brakes. A few tenders were equipped with a back cab for tender-first running in their latter days when the engines were employed mainly on branch services.

In 1946, the surviving members of the class were renumbered 2225 - 2230 (ex-6878–6883), 2231–2244 (ex-6819–6824 and 6893–6900) and 2245–2256 (ex-6901–6912). Two had been withdrawn in 1946 – 6821 and 6879 – and were not renumbered. More were withdrawn in 1947 and twenty-two became BR stock in 1948, numbered 62225–62256, with gaps for the withdrawn engines that had been allocated numbers. Most were repainted in BR livery with BRITISH RAILWAYS on the tender rather than the later lion & wheel emblem, and 62234, 62238, 62240 and 62249 did not carry the smokebox door numberplate. The three last survivors - 62225, 62241 and 62242 - were withdrawn in February 1953.

2243, formerly GNoS 99 built in December 1895, seen here c1948. It was withdrawn as BR 62243 in 1951. (Photomatic. MLS Collection)

62225 of 1893 and 62246 of 1897 at Elgin, 15 June 1951. 62246 was withdrawn a couple of months later but 62225, the 'S' prototype was one of the last survivors, withdrawn in 1953. (J.D. Darby/MLS Collection)

136 • NORTH EASTERN, NORTH BRITISH, GREAT NORTH OF SCOTLAND, L N E R

Another pair, the first and last, 62256 (the former 112 of 1898) and 62225 again, on shed at Keith at the end of the day, 17 June 1951. (J.D. Darby/MLS Collection)

Below left: One of the three last survivors withdrawn in February 1953, 62241, at Huntley shortly before withdrawal. Burning of the smokebox door was clearly common on the class of locomotive. (MLS Collection)

Below right: **62248 at** Keith just two months before its withdrawal, August 1952. (Photomatic/John Scott-Morgan's Collection)

Operation

Initially the class was based at Kittybrewster, Keith and Elgin for main line work between those locations, although the main express turns were mostly covered by the Pickersgill 'V' and 'F' classes from the first decade of the twentieth century.

I can trace just one log of a D41 on other than an all-stations stopping or branch train. It was timed by O.S. Nock and dates from the mid-1930s.

Keith–Aberdeen

6907, 210 tons

3.30pm Inverness–Aberdeen

Miles	Location	Times	Speed		Gradients
0	Keith	00.00			
4.3	Grange	07.00	53		L
5.1	Cairnie Junction	07.48	58½		1/200 F
8	Rothiemay	10.55	46½		1/160 R
12.5	Huntley	17.00		1 E	
0		00.00		T	
5	Gartly	09.15	46		1/210 R
8	Kennethmont	13.35	35½		1/100 R
9.7	Wardhouse	16.00	53		1/142 F
13.2	Insch	19.35	63½		1/200 F
16.2	Oyne	23.05	46/60		1/550 R, 1/100 F
19.4	Pitcaple	26.30	53/56		1/140 R, 1/100 F
20.4	Inveramsay	28.00		1 E	(Conditional stop)
0		00.00		1 E	
	MP 17	-	51		1/300 F
3.5	Inverurie	05.50		¼ E	
0		00.00		T	
3.5	Kintore	06.10	52½		L
6.3	Kinaldie	09.15	56		L
8.6	Pitmedden	10.40	58½		L
10.6	Dyce	14.15		¼ L	
0		00.00		T	
2	Bucksburn	04.15			
3.7	Woodside	06.00	53		1/275 F
4.9	Kittybrewster	07.50	eased		1/150 F
5.9	Aberdeen	09.40		¼ E	

107 on a passenger train (note, no lamp headcode) at an unknown location, c1905. (MLS Collection)

2249 on the 3pm from Craigellachie at Boat of Garten, 19 June 1947. (W.A. Camwell/MLS Collection)

By the introduction of the final raft of Heywood 'Fs' in 1920, the Johnson engines were distributed widely covering both passenger and freight work.

In 1935 their allocation was:

Elgin:	6819, 6881, 6894, 6905
Keith:	6820-6823, 6878-6880, 6882, 6893, 6896, 6899, 6909
Kittybrewster:	6824, 6883, 6895, 6897, 6898, 6900-6904, 6906-6908, 6910-6912.

During the Second World War, freight traffic in Scotland increased considerably and four were transferred to the former NB section to assist with the additional freight work. 6882 and 6883 moved to Fife (Thornton Junction and Dunfermline) and 6880 and 6898 to St Margaret's although they were found insufficiently powerful for most of that depot's diagrams, being used mostly on local pick-up goods work. In 1943 all four returned to the GNoS section.

In 1946, 6894-6896, and 6899-6912 were based at Keith and the remainder at Kittybrewster. After nationalisation they were used mainly on branch line work as well as the Speyside lines to Boat of Garten, the Thompson B1s having displaced all the former GNoS 4-4-0s from express work by this time. Their last duties included station pilot work at Aberdeen and Elgin and the Lossiemouth branch.

62230 (the former 'S' 83 of 1893) at Fraserburgh with the 9.37am from Maud, 19 July 1949. (J.D. Darby/MLS Collection)

2256 on a southbound goods train at Inverurie, 20 July 1949. (J.D. Darby/MLS Collection)

62225, the oldest D41 and last survivor, entering Lossiemouth with the 5.10pm from Elgin, 15 June 1951. (J.D. Darby/MLS Collection)

62255 leaving Buckie with a goods train for Cullen, 15 June 1951. (J.D. Darby/MLS Collection)

The Great North of Scotland 'O' Class (LNER D42)

In 1888, there was a need for larger engines than the earlier Cowan and Manson 4-4-0s as passenger traffic increased and at the behest of the directors, Manson designed the 'O' class with larger firebox, increased cylinder diameter and 6ft coupled wheel diameter. Nine locomotives were delivered by the Kitson Company in 1888 and they were numbered 4, 7, 9, 10, 17 and 18 (revenue account) and 72–74 (capital account).

Their dimensions were:

Cylinders (2-inside):	18 x 26in
Coupled wheel diameter:	6ft 0½in
Bogie wheel diameter:	3ft 9½in
Boiler pressure:	165lbs psi
Stephenson motion with slide valves	
Heating surface:	1,165sq ft
Grate area:	18.2sq ft
Axleload:	15 tons 3 cwt
Weight (Engine):	44 tons
(Tender):	29 tons
(Total):	73 tons
Water capacity:	2,100 gallons
Coal capacity:	3 tons (later 4 ½ tons)
Tractive effort:	16,296lbs

A decision had been made to change the diameter of the bogie wheels from the previous 3ft to 3ft 9½in, at an extra cost of £25 added to the contract price of £1,950, and such large bogie wheels remained the standard for subsequent GNoS designs.

During routine reboilering between 1915 and 1920, five were equipped with superheaters, 74 with the Schmidt type and 17, 18, 72 and 73 with the Robinson equipment. 74 was subsequently fitted with the Robinson type. The overall heating surface of the superheated boilers was reduced to 1,000sq ft, but this included a superheated surface of 140sq ft. The boiler pressure was reduced to 160lbs psi, reducing the theoretical tractive effort to 15,802lbs, but increasing the axleload by 15 cwt and engine weight by 2 tons 7 cwt. The superheated engines had an extended smokebox and all had their GNoS smokebox doors replaced by the NB type, as with other 4-4-0 classes. After the experiment with eight-wheel tenders for the Manson 'Q' class, the 'O' class reverted to six-wheel tenders with limited water and coal capacity, but the addition of coal rails before the Grouping increased the tender capacity by a further 1½ tons of coal.

At the Grouping, the nine locomotives were given the classification D42 and were renumbered 6804, 6807, 6809, 6810, 6817, 6818 and 6872-6874. 6804 was withdrawn in 1935 and the rest, apart from 6807 and 6817, were withdrawn before the Second World War. The additional military traffic requirements stayed the execution of this pair which were withdrawn at the end of the war, 6807 in April 1945 and 6817 in February 1946. Both had been allocated the 1946 LNER numbers of 2075 and 2076 but neither were carried.

Manson's Class 'O' No.10 built in April 1888, seen in the first decade of the twentieth century. (MLS Collection)

Right: **6810 with** NB smokebox door, wingplates removed, and tender cab board, in LNER black livery, c1935. (F. Moore/MLS Collection)

Far right: **6874 also** with tender cab, seen at Elgin, 9 June 1936. (MLS Collection)

6807, built in 1888, renumbered and painted in LNER lined green livery, June 1924. (MLS Collection)

Operation

Like the other GNoS 4-4-0s, they commenced duties on main line Aberdeen–Keith and Elgin expresses but were cascaded onto lower priority services as the 'S', 'T', 'V' and 'F' 4-4-0s were introduced. In the 1920s, 6804 was based at Elgin for the Lossiemouth branch and 6807 was sub-shedded at Boat of Garten. The remainder were at Kittybrewster. 6809 was at Kittybrewster's sub-depot of Macduff and 6818 at Banchory. 6817 moved to Alford in the 1930s. Their final allocations from which they were withdrawn in the 1930s was:

Keith: 6807*, 6872, 6873
Elgin: 6874
Kittybrewster: 6804, 6809, 6810, 6817*, 6818

(* not withdrawn until 1945/6).

The Great North of Scotland 4-4-0s • 143

6810 with an Elgin train at the Lossiemouth terminus, c1935. (Photomatic/John Scott-Morgan Collection)

6874 at Elgin with the branch train for Lossiemouth, 11 August 1937. (N. Fields/MLS Collection)

Great North of Scotland 'P' Class (LNER D43)

This class consisted of just three locomotives, part order of six built by Robert Stephenson & Co. in 1890. The other three were described previously, Class 'Q' (LNER D38). They had 6ft 6in coupled wheels, these had 6ft, otherwise they were identical. They were delivered in May and June 1890, numbered 12–14, and classified 'P'. Their dimensions were:

Cylinders (2-inside):	18 x 26in
Coupled wheel diameter:	6ft 0½in
Bogie wheel diameter:	3ft 9½in
Boiler pressure:	165lbs psi
	Stephenson motion with slide valves
Heating surface:	1,165sq ft
Grate area:	18.2sq ft
Axleload:	15 tons 3 cwt
Weight (Engine):	44 tons
Weight (Tender):	37 tons 8 cwt
Weight (Total):	81 tons 8 cwt
Water capacity:	3,000 gallons
Coal capacity:	3 tons (later 5 tons)
Tractive effort:	16,296lbs (15,802lbs for superheated engines)

12 and 14 received superheated boilers in 1916/7, both reverting to saturated steam at boiler changes, then 6814 acquiring a superheated boiler again in the 1930s. The superheated boilers had a heating surface of 1,000sq ft including 140sq ft of superheating, reduced boiler pressure of 160lbs psi and weight increase of nearly 2½ tons. The superheated boilers had an extended smokebox and were without smokebox wingplates. Like the 6ft 6in 'Qs' (D38), they were equipped with similar eight-wheel tenders, free running bogie at the front, rigid framed axles at the rear. These were replaced with standard six-wheel tenders before the Grouping for 12 and 14, and 6813 after. As indicated, they were renumbered in 6812–6814 at the Grouping and classified D43.

All were withdrawn before the Second World War, 6814 in 1936, 6813 in 1937 and 6812 in 1938.

6813, former GNoS Class 'P', LNER D43, built in 1890 as No.13, still with GNoS smokebox door and wingplates, in LNER lined green livery, November 1924. (MLS Collection)

The Great North of Scotland 4-4-0s • 145

6814 in LNER black livery, smokebox door replaced and wingplates removed, at Banff, 9 June 1936. (MLS Collection)

6813 with cab tender board, at Kittybrewster, c1937 just before withdrawal. (F. Moore/MLS Collection)

Operation

All three were allocated to Kittybrewster initially for Aberdeen–Elgin passenger trains. 13 moved to Elgin before the Grouping but returned to Kittybrewster, before finishing its days at Elgin. 6812 finished at Elgin, 6814 at Kittybrewster, both used entirely on branch line services.

Great North of Scotland 'A' Class (LNER D44)

Back to 1884 and Manson's first class of 4-4-0s – six locomotives numbered 63–68 constructed by Kitson & Co. They had been ordered in 1881 in Cowan's regime, but Manson came in 1883 and made a number of design changes, the most noticeable being the placing of the inside cylinders and the position of the dome on the boiler barrel instead of on the firebox. They were classified 'A' and their key dimensions were;

Cylinders (2-inside):	17½ x 26in
Coupled wheel diameter:	6ft 0in
Bogie wheel diameter:	3ft 0in
Boiler pressure:	140lbs psi

Stephenson motion with slide valves

Heating surface:	1,036sq ft
Grate area:	18sq ft
Axleload:	12 tons 2 cwt
Weight (Engine):	37 tons 2 cwt
(Tender):	29 tons
(Total):	66 tons 2 cwt
Water capacity:	2,000 gallons
Coal capacity:	3 tons

Four of the class, 63, 64, 67 and 68 were rebuilt with larger boilers in 1905/6 and the other two, 65 and 66, in 1912. This increased the boiler pressure to 150lbs psi, the heating surface to 1,1144sq ft, decreased the grate area to 16.05sq ft and increased the engine weight by 4½ tons and the maximum axleload by over two tons.

They retained their GNoS style smokebox doors until withdrawal but their smokebox wingplates were removed when they were reboilered. They were fitted with Westinghouse air brakes only. They had six-wheel tenders, but at the Grouping, 66 and 67 were fitted with small four-wheel tenders carrying just 2 tons of coal and 1,050 gallons of water to enable the engines to fit the turntables at Kintore and Alford. These tenders were swapped with other members of the class. 65 had a tender cab for a time.

They were allotted 6863–6868 at the Grouping although only 6865 and 6867 ever carried the LNER numbers. They were classified D44. 6867 reverted to a six-wheel tender with coal rail and back cab increasing the coal capacity to 4½ tons. 63, 64, 66 and 68 were withdrawn in 1924/25, 6863 in August 1926, but 6867 lingered on to October 1932.

Operation

They operated for years from Kittybrewster with both passenger and freight on the Buchan lines, often outbased to Fraserburgh or Peterhead. At the Grouping 63 and 65 performed shunting and

One of the two D44s with the four-wheel tender, 67, at Kittybrewster, c1923. It was also fitted with a back cab board. (Real photographs/MLS Collection)

local duties at Elgin, 64 was sub-shedded at Meldrum for the branch to Inverurie and 66 and 67 with their small tenders worked from Alford to Kintore. 68 and 6867 in its later years worked mainly on engineering trains and shunting turns from Kittybrewster.

Great North of Scotland Class 'M' (LNER D45)

Cowan's last 4-4-0s were his class 'M' of 1878, nine engines with 5ft 7in coupled wheels built by Neilson & Co., part of an order for twelve locomotives, the other three engines having 6ft 1in coupled wheels (GNoS Class 'C', LNER D39). They were enlarged versions of his earlier class 'L'. they were numbered 40, 51, 53, and 57–62. Their dimensions were:

Cowan's class 'M' No.51, built in 1878, rebuilt with a Pickersgill boiler in 1899, seen at Kittybrewster, c1920. (Real photographs/MLS Collection)

Cylinders (2-outside):	17½ x 26in
Coupled wheel diameter:	5ft 7in
Bogie wheel diameter:	3ft 0½in
Boiler pressure:	140lbs psi

Stephenson motion with slide valves

Heating surface:	1,107.4sq ft
Grate area:	14sq ft
Axleload:	13 ¾ tons
Weight (Engine):	41 tons
(Tender):	29 tons
(Total):	70 tons
Water capacity:	1,950 gallons
Coal capacity:	3 tons (4 tons when coal rails added around 1920-25)

Rebuilding with larger boilers was undertaken by Pickersgill between 1896 and 1904. The new boilers had a pressure of 150lbs psi, heating surface of 1,118.5sq ft, grate area of 15.64sq ft and with a maximum axleload of 14 tons 1 cwt, weighed 42 tons 4 cwt. The boilers for 53, 58 and 62 were supplied by Kitson's and the rest were rebuilt at Kittybrewster.

At the Grouping, the nine locomotives were classified LNER class D45 and were allocated numbers 6840, 6851, 6853 and 6857–6862. Nos. 57, 60 and 62 were withdrawn before receiving their new numbers and all had gone by 1927 apart from 6840, which like 6867 of the D44 class, lingered on to 1932.

Operation

After initial work on the Aberdeen–Huntley–Keith–Elgin main line on both passenger and freight work, they were relegated to local and branch services as the Manson 4-4-0s were introduced in the late 1880s. After the First World War, 58 and 62 were based at Elgin, the rest being at Kittybrewster. After the Grouping a number resided at the branch sub-sheds – 6840 and 6859 at MacDuff, and 6851, 6858 and 6861 at Keith. 6840 retained a role as spare engine at Kittybrewster until its withdrawal in 1932.

Great North of Scotland Class 'N' (LNER D46)

Unlike previous locomotive construction which was undertaken until the mid-1880s by contracting engineering companies, the Kittybrewster Works was capable of erecting two locomotives from parts purchased from various firms. They were 5ft 7in coupled wheel

No.5 at the Grouping with LNER livery but as yet unrenumbered, c1923. (F. Moore/MLS)

No.6 at Kittybrewster with a stopping train to Dyce, September 1923. No.6 is in LNER livery but not yet renumbered and has retained its smokebox wingplates. (RCTS Collection)

machines and were completed in 1887 to Manson's design, classified 'N' and numbered 5 and 6. Unusually for the GNoS at that time they were named, No.5 as *Kinmundy,* the name of the company Chairman's home and how he was generally known and No.6 *Thomas Adam* after the Vice-Chairman. Their dimensions were:

Cylinders (2-inside):	17½ x 26in
Coupled wheel diameter:	5ft 7in
Bogie wheel diameter:	3ft 1in
Boiler pressure:	140lbs psi

Stephenson motion with slide valves

Heating surface:	1,186.7sq ft
Grate area:	16.51 st ft
Axleload:	13½ tons
Weight (Engine):	40 tons
(Tender):	26 tons
(Total):	66 tons
Water capacity:	2,100 tons
Coal capacity:	3 tons

Like previous and succeeding 4-4-0s, they were rebuilt with new boilers with higher pressure, 165lbs psi, heating surface of 1,159sq ft and weighed 42 tons 5 cwt. Their tenders were fitted with coal rails increasing the capacity to 4½ tons and their weight to 29 tons and both had tender cabs. The boiler replacement took place in 1915 (No.5) and 1917 (No.6) and they were renumbered 6805 and 6806 at the Grouping and classified D46. Despite the reboilering both retained the GNoS style smokebox door and No.6 retained its smokebox wingplates although No.5's were removed before the Grouping. Pickersgill had removed the names of the two engines when they went through Works in the last decade of the nineteenth century. Both somewhat surprisingly lasted into the 1930s, 6806 being withdrawn in 1932 and 6805 in 1936.

Operation

Both locomotives were based at Kittybrewster and were used on passenger and goods work and latterly on engineering trains, local goods and shunting. 6805 was used for at time at Alford for the Kintore branch.

Great North of Scotland Classes 'K' & 'L' (LNER D47)

We move back to the earliest era of the GNoS 4-4-0s. The very earliest, Cowan's class 'H', was extinct before the Grouping and therefore does not qualify for inclusion in this book, but the class 'K' of 1866 does, by virtue of the fact that three of the six engines lasted until 1925. They were built by Neilson & Co. and were numbered 43–48. Their dimensions were:

Cylinders (2-outside):	16 x 24in
Coupled wheel diameter:	5ft 6½in
Bogie wheel diameter:	3ft 0in
Boiler pressure:	140lbs psi
Stephenson motion with slide valves	
Heating surface:	1,046sq ft
Grate area:	14.07sq ft
Axleload:	12 tons 5 cwt
Weight (Engine):	37 tons 5 cwt
(Tender):	27 tons
(Total):	64 tons 5 cwt
Water capacity:	1,800 gallons
Coal capacity:	4 tons
Tractive effort:	10,994lbs

They were rebuilt with slightly larger boilers between 1889 and 1891 and their simple weatherboards replaced by a cab. Nos. 43, 46 and 47 were withdrawn in 1921 and the three survivors had been renumbered in the duplicate list as 44A, 45A and 48A between 1916 and 1921, but were withdrawn in 1925 before LNER renumbering. They were however belatedly given the classification D47/2 to distinguish them from

Above: **'K' No. 45** as rebuilt in 1891 with slightly larger boiler and provision of cab. Note the copper capped chimney retained until withdrawal, brass dome, outside cylinders and 4-wheel tender, which could contain only 1,050 gallons of water and 2 tons of coal. It is seen here c1900. (MLS Collection)

Below: **'K' No. 43A,** built in 1866 and rebuilt with a slightly larger boiler in 1890, renumbered on the duplicate list in 1916, seen here with six-wheel tender and retaining its tall copper capped chimney

150 • NORTH EASTERN, NORTH BRITISH, GREAT NORTH OF SCOTLAND, L N E R

Another shot of 'K' No.45 in GNoS days, but on the duplicate list as 45A. It still has the 4-wheel tender but a back-board has been provided to give protection when running tender-first. It is seen c1920. (Real Photographs/MLS Collection)

D47/2 (former 'K') No.48A after withdrawal at Inverurie Works, August 1925. It has a 6-wheel cab-tender. It has Westinghouse equipment for both engine and train and an extra pump for bridge-riveting apparatus. (MLS Collection)

the later Cowan class 'L' engines. They retained tall copper capped chimneys to the end.

The class 'L' were of similar design, six engines being built in 1876, over-ordered by Cowan in 1873 as only four had been authorised. They were numbered 49, 50, and 54–57, although 57 was later renumbered 52. Cowan had ordered some improvements over the class 'K', namely:

Cylinders	
(2-inside):	17 x 24in
Boiler pressure:	150lbs psi
Heating surface:	1,118.5sq ft
Grate area:	15.64sq ft
Axleload:	14 tons 7 cwt
Weight (Engine):	41 tons 3 cwt
(Tender):	28 tons
(Total):	69 tons 3 cwt
Tractive effort:	13,298lbs

Rebuilding with new boilers took place between 1897 and 1901. No.52 was rebuilt after being involved in a collision and was the only one to receive a side-window cab. All six were taken into LNER stock, still with GNoS duplicate numbers, the 'A' added to their original numbers. They were all withdrawn in 1924, surprisingly before the three remaining engines of class 'K', apart from 52A (the original 57) which was withdrawn in 1926. In their short LNER careers, they were classified D47/1.

Operation

These engines were treated as mixed traffic machines from the outset and were employed generally throughout the system on both the main line and branches. 48 and 52 spent time on the Buchan section, 49, 54 and 55 on the Deeside line, 44 on the Speyside and 47 on the Lossiemouth branch. 45 was the regular branch engine at Old Meldrum. By the Grouping all were based at Kittybrewster and used on light local work.

In 1925, 45A was repainted in GNoS green livery and displayed at the Darlington Centenary celebrations with the possibility of preservation but after a spell as shunter at Inverurie Works, it was withdrawn in July 1925 and scrapped.

The Great North of Scotland Class 'G' (LNER D48)

Three 4-4-0 locomotives with 5ft 6in coupled wheels were part of an order placed with Kitson & Co. in 1881 – the rest consisting of the Class 'A' and 'N' 4-4-0s and 0-6-0 tanks of classes 'D' and 'E'. They were goods engines, the 'As' having 6ft coupled wheels. They were delivered in 1885 and numbered 69–71 and were classified 'G'. Their dimensions were:

Cylinders (2-inside):	17½ x 26in
Coupled wheel diameter:	5ft 6in
Bogie wheel diameter:	3ft 0in
Boiler pressure:	140lbs psi

Stephenson motion with slide valves

Heating surface:	1.036sq ft
Grate area:	18sq ft
Axleload:	11 tons 16 cwt
Weight (Engine):	36 tons 10 cwt
(Tender):	23 tons 11 cwt
(Total):	60 tons 1 cwt
Water capacity:	2,000 gallons
Coal capacity:	3 tons (4 tons later when coal rails fitted)

They were rebuilt with new boilers between 1905 and 1912. Boiler pressure was increased to 150lbs psi, heating surface to 1,144sq ft and the grate area reduced to 16.05sq ft. The weight of the engine increased to 41 tons 6 cwt, the tender to 27 tons 12 cwt and the axleload to 14 tons 7 cwt. They retained their GNoS smokebox doors but lost their wingplates before the Grouping. All three had tender boards fitted to protect the crews when running tender-first. At the Grouping, they were renumbered 6869–6871 and were classified D48. 6870 and 6871 were withdrawn in 1928 and 6869 survived until November 1934.

Operation

Their main work was fish traffic on the Buchan section in the far north-east, but they also worked passenger trains on the main line and branches. After the Grouping, 6869 and 6870 were based at Kittybrewster, 6870 sub-shedded at MacDuff. 6871 was allocated to Keith and sub-shedded at Banff for the Tillynaught branch. After the withdrawal of the first two, 6869 remained at Kittybrewster as a spare engine.

D48, former GNoS class 'G', 6871 at Cornhill on a stopping train, August 1925. 6871 has a tender with coal rails and a backing board/cab extension. (John Scott-Morgan Collection)

Chapter 5
THE LNER D49 'SHIRE' & 'HUNT' 4-4-0s

The LNER inherited 920 4-4-0s at the Grouping, with well over 300 of them built in the nineteenth century and the majority of the rest dating from the first decade of the twentieth. The LNER's new Chief Engineer, the Great Northern's Nigel Gresley, had just introduced his 'A1' Pacific locomotive *Great Northern* and the third of the design, *Flying Scotsman*, was shortly to be exhibited at the 1924 Wembley Exhibition. His priority was to provide the East Coast Main Line with more of these powerful Pacifics to cope with the increasing train loads that in the 1920s were taxing the haulage capabilities of the excellent Great Northern and North Eastern Atlantics and 4-4-0s. The only immediate need for a new design for secondary services was in Scotland where some twenty-four locomotives of the North British D26–D28 and D35s were life expired and Gresley selected the Great Central 'D11' 'Director' 4-4-0s to fill the gap as was documented in my earlier volume of the LNER 4-4-0s which covered the Great Central design.

By 1925, there were fifty-two Pacifics in service and Gresley then turned to the needs of secondary services, particularly those in the North East and Scotland, the areas with the oldest numbers of pre-Grouping 4-4-0s, the Great Central D10s and D11s and the Great Eastern D15s being comparatively younger and still capable of fulfilling their line duties. There were areas, especially in Scotland where the axleload of the Pacifics was too high – the Waverley route and north of Dundee to Aberdeen being the most important of these routes. Gresley therefore specified to his design team in Darlington the need for a 3-cylinder 4-4-0 and the design was completed by early 1926. The building of twenty-eight machines to this design was authorised by the LNER Board in April 1926 and twenty of these were constructed at Darlington Works in 1927 and 1928. These twenty had the following key dimensions:

Cylinders (3):	17 x 26in
Coupled wheel diameter:	6ft 8in
Bogie wheel diameter:	3ft 1¼in
Boiler pressure:	180lbs psi
	Walschaerts/Gresley motion with piston valves
Heating surface:	1,669.58sq ft (incl 271.8sq ft of superheating)
Grate area:	26sq ft
Axleload:	21 tons 5 cwt
Weight (Engine):	65 tons 11 cwt
(Tender – LNER):	52 tons 13 cwt
(Tender – ex NER):	44 tons 2 cwt
(Tender – ex GCR):	47 tons 6 cwt
Water capacity:	4,200 gallons (LNER); 4,125 gallons (NER); 4,000 gallons (GCR)
Coal capacity:	7½ tons (LNER); 5 ½ tons (NER); 6 tons (GCR)
Tractive effort:	21,556lbs

The locomotives were funded from the revenue account replacing withdrawn locomotives so the numbers allocated were random gaps until the 1946 renumbering scheme rationalised the system. The first twenty were named after counties

The first D49, 234 *Yorkshire*, as built in 1927 in Works photographic grey livery.
(Locomotive Publishing Co./MLS Collection)

through which the LNER passed and were known as class D49, the 'Shires'.

234	*Yorkshire*
251	*Derbyshire*
253	*Oxfordshire*
256	*Hertfordshire*
264	*Stirlingshire*
265	*Lanarkshire*
266	*Forfarshire*
236	*Lancashire*
270	*Argyllshire*
277	*Berwickshire*
245	*Lincolnshire*
281	*Dumbartonshire*
246	*Morayshire*
249	*Aberdeenshire*
250	*Perthshire*
306	*Roxburghshire*
307	*Kincardineshire*
309	*Banffshire*
310	*Kinross-shire*
311	*Peebles-shire*

In 1928, as part of the same order, Darlington built six 4-4-0s with oscillating cam operated Lentz poppet valve gear. They were numbered and named:

318	*Cambridgeshire*
320	*Warwickshire*
322	*Huntingdonshire*

245 *Lincolnshire*, built in February 1928, at Doncaster in LNER lined green livery.
(F.Moore/MLS Collection)

266 *Forfarshire* at Dundee, c1930.
(MLS Collection)

327 *Nottinghamshire* as built in 1928 with oscillating cam Lentz valve gear and Westinghouse brake. (MLS Collection)

336 *Buckinghamshire* built in 1929 with rotary cam Lentz poppet valve gear on test at Cowlairs. When further examples of this variety were built and named after 'Hunts', 336 was renamed *The Quorn.* (MLS Collection)

327 *Nottinghamshire*
329 *Inverness-shire*
335 *Bedfordshire*

The final pair of the initial order were built with rotary cam operated Lentz poppet valve gear. They were numbered and named:

336 *Buckinghamshire*
352 *Leicestershire*

A further order for eight piston valve engines was made in 1928 and 2753–2760 were delivered in 1929 and were named:

2753 *Cheshire*
2754 *Rutlandshire*
2755 *Berkshire*
2756 *Selkirkshire*
2757 *Dumfries-shire*
2758 *Northumberland*
2759 *Cumberland*
2760 *Westmorland*

Above: 2757 *Dumfriess-shire* as built in 1929 with a straight-sided tender, c1930. (MLS Collection)

The six locomotives with oscillating poppet valve gear had problems, unlike the rotary cam valve gear engines and a new order for a further fifteen engines in 1929 were placed for 4-4-0s with the rotary gear, and these were classified D49/2. The six with the oscillating gear were rebuilt in 1938 with piston valves as class D49/1, the same as the earlier piston valve 'Shires'. They, like the earlier 'Shires' were equipped with LNER standard tenders with stepped-out side panels. The 1929 ordered engines were not delivered from Darlington Works until 1932 and 1933 and were numbered and named:

Below: The first of the 1932 built D49/2s, 201 *The Bramham Moor*, in photographic grey, as built. (F. Moore/MLS Collection)

201 *The Bramham Moor*
211 *The York and Ainsty*
220 *The Zetland*

232	*The Badsworth*	282	*The Hurworth*
235	*The Bedale*	283	*The Middleton*
247	*The Blankney*	288	*The Percy*
255	*The Braes of Derwent*	292	*The Southwold*
269	*The Cleveland*	297	*The Cottesmore*
273	*The Holderness*	298	*The Pytchley*

Then, in October 1933, a final order for twenty-five more D49/2s was made, and they were constructed at Darlington in 1934 and 1935, numbered and named:

205	*The Albrighton*
214	*The Atherstone*
217	*The Belvoir*
222	*The Berkeley*
226	*The Bilsdale*
230	*The Brocklesby*
238	*The Burton*
258	*The Cattistock*
274	*The Craven*
279	*The Cotswold*
353	*The Derwent*
357	*The Fernie*
359	*The Fitzwilliam*
361	*The Garth*
362	*The Goathland*
363	*The Grafton*
364	*The Grove*

235 *The Bedale* newly built in LNER lined green livery (having avoided the decision in 1928 to paint all pre-grouping 4-4-0s black), at York in 1934. (Photomatic/MLS Collection)

247 *The Blankney* at Scarborough in 1937. (E.R. Morrten/MLS Collection)

365 *The Morpeth*
366 *The Oakley*
368 *The Puckeridge*
370 *The Rufford*
374 *The Sinnington*
375 *The South Durham*
376 *The Staintondale*
377 *The Tynedale*

Finally, as mentioned earlier the two Rotary cam valve engines built experimentally in 1929, were renamed as members of the 'Hunt' D49/2 class:

336 *The Quorn*
352 *The Meynell*

The variant dimensions of the D49/2s were:

Lentz rotary cam poppet valves
Axleload: 21 tons
Weight (Engine):   64 tons 10 cwt

The first large batch of D49/2s had five fixed cut-off positions in the forward gear and this was considered too restrictive in operation. 282 *The Hurworth* delivered in 1932 was modified to give an infinite variation of cut-off positions between 15% and 84%, but this was removed at 282's first general repair in 1934. It emerged from Darlington then with seven cut-off positions – 15, 20, 25, 35, 45, 60 and 78%. One of the earlier D49/2s, 269 *The Cleveland*, was similarly altered in 1934 and the twenty-five 1934/35 built engines all had this cut-off arrangement, but the remained of the 1932/33 engines remained with the five fixed positions.

Comparative tests were run in the autumn of 1935 and early 1936 between D49/2s, 292 *The Southwold* (five cut-off positions) and 377 *The Tynedale* (seven cut-off positions) and D49/1 piston valve 251 *Derbyshire*. The tests were exhaustive at varying speeds and cut-off positions using the ex NER counter-pressure locomotive and the results were unexpected and disappointing in that the D49/1 showed a marked superiority in economy to undertake the same outputs. As a result, the six oscillation cam poppet valve engines were converted in 1938 to piston valve D49/1s rather than in conformity with the rotary cam valve D49/2s. It is possible that the entire class of D49/2 'Hunts' might have been rebuilt as piston valve engines but for the interruption caused by the onset of the Second World War.

The tenders used for the first twenty-eight 1927-1929 constructed locomotives were the LNER standard 4,200 gallon ones with stepped-out side panels as fitted also to the K3 moguls and J38 0-6-0s. However, the eight D49/1s, 2753–2769 were fitted to LNER standard tenders with straight sides. However, it was realised that the J38s did not need the water and coal capacity for their work in Scotland and fifteen of their tenders with the stepped-out side panels were transferred to the 1932 built 'Hunts'. Then the final twenty-five 1934/35 D49/2s were provided with the straight-sided tenders.

By 1938, all had LNER standard tenders of the two outlines conveying 4,200 gallons of water, 7½ tons of coal and water pick-up scoops. However, few of the class had turns which required picking up water en route and when the D49s with the oscillating cam valves were rebuilt their tenders were used for new V2 2-6-2s and they were equipped with NER designed tenders from Q6 0-8-0s with 4,125 gallon capacity, 5½ tons of coal space and no water scoop. Then, in 1941 tenders became available from ex GC Q4 0-8-0s that were being rebuilt into heavy shunting tank engines and twenty-eight D49/1s exchanged their standard LNER tenders for these, enabling theirs to be fitted to new O2 2-8-0s under construction at Doncaster, saving precious steel. These tenders could contain 4,000 gallons of water and 6 tons of coal.

The D49s had Wakefield mechanical lubricators of varying types as efforts were made to get satisfactory lubrication, after some early problems with hot boxes. Gresley 3-cylinder engines were also prone to overheating of the middle big-end bearing and a fluid container was fitted inside the crank axle which when overheating occurred produced a gas smelling of violets, later changed to aniseed to alert the driver. 282 was the first D49 so equipped in 1939 and the rest were fitted as they went through Works overhauls.

Some problems with poor steaming were reported in the early days and Gresley was interested in the French experience with the Kylchap exhaust system used by Chapelon so successfully (and on five of Gresley's A4s). 251 *Derbyshire* of Neville Hill, a poor steamer, was selected and emerged with the Kylchap blastpipe in October 1928. After slight adjustment of blastpipe, petticoat and chimney, results were good and 322 *Huntingdonshire*, a D49 with oscillating cam poppet valves, was fitted in April 1922. Despite the

The Thompson 1942 rebuild, class 'D' 365 *The Morpeth*, at Neville Hill, c1945. (MLS Collection)

The 'D' in BR days, a few months before its withdrawal, renumbered 62768 and in BR lined black, at Starbeck, 5 October 1951. (N. Harrop/MLS Collection)

Kylchap success elsewhere, it was concluded in 1930 that the Kylchap exhaust on the two D49s had made insufficient difference to warrant the expense and it was removed from both engines.

When built, all received the full LNER lined green livery and were not subject to the edict of 1928 that the 4-4-0 fleet should be repainted black with red lining as an economy measure. However, from 1941 they were painted plain black as the war took its toll and although many classes regained lined green after the war, no D49s were so treated and their black livery was exchanged after nationalisation for the BR lined black mixed traffic livery. In the 1946 renumbering scheme, they were allocated 2700–2775 and in 1948 62700–62775.

In 1942, the only subsequent rebuilding of any D49 took place as part of Edward Thompson's programme of replacing Gresley 3-cylinder engines with two. 365 *The Morpeth* was selected as it had been experimentally fitted in the autumn of 1939 with infinitely variable rotary cam poppet valve gear but removed in 1941 because of damage to the gear. It emerged in August 1942 with two inside cylinders of the GC 'Director' pattern, 18 x 26in, with Stephenson valve gear and piston valves, reducing its tractive effort to 19,890lbs. Its performance was undistinguished and no more were rebuilt and this rather ungainly engine was withdrawn in 1952 following collision damage. It had been given the simple classification 'D' rather than a subgroup of the D49s. Surprisingly the damaged rotary cam valve gear was repaired and fitted in 1949 to 62764 *The Garth* with some modifications in accordance with the Reidinger gear patents and the engine was tested on Rugby Testing Station and worked successfully subsequently from Neville Hill depot retaining the gear until its withdrawal in November 1958. However, no other D49/2 was modified in this way.

The D49s had been criticised for their rough riding from the beginning and efforts were made to improve the springing. 253 was fitted with D11 'Director' springs with little improvement. Gresley discussed the problem with Stanier (then still at Swindon) and 329 *Inverness-shire* was fitted with 15-plate springs ½in thick and this appeared to improve the riding, so

The last days – 62768 at Holbeck, 26 July 1952, just weeks before its collision resulting in its demise. (N. Fields/MLS Collection)

all the existing D49s in 1929 were so equipped including the first fifteen D49/2s. However, broken springs were still common and at the end of 1933 further changes were made, with 2759 *Cumberland* receiving sixteen-leaf 5in wide and ½in thick springs. All fifty-one engines were so modified by 1935 and the last constructed engines were fitted with them during their building. However, because of ongoing complaints of continued rough-riding after the war, five D49/2s, 62738 and 62741–62744, were fitted with cast steel axleboxes with manganese steel liners and redesigned bogies similar to the Thompson B1s. Three were retained in the North East, but 62743 and 62744 went permanently to Scotland.

Six were withdrawn in 1957 – one D49/1 and five D49/2s – and twenty-nine went to the scrapyard in 1958. Withdrawals followed rapidly and at the beginning of 1961 only eight D49/1s and four D49/2s were surviving. All had gone by the end of the year, the last being 62712 *Morayshire* in July, which was subsequently purchased for preservation (see Chapter 7).

D49/1 62729 *Rutlandshire*, built in 1929, in BR lined black livery and with straight-sided LNER standard tender, at Thornton Junction, 7 October 1948. (J.D. Darby/MLS Collection)

D49/1 62705 *Lanarkshire* with ex-GC tender, at Haymarket, September 1955. (Photomatic/. MLS Collection)

Below left: **62728** *Cheshire*, with LNER straight-sided tender, at Dundee Tay Bridge, September 1955. (MLS Collection)

Below right: **62716** *Kincardineshire* with ex-NER tender from a Q6, at Thornton Junction, 4 August 1957. This was one of the last D49 survivors, withdrawn in April 1961. (MLS Collection)

62742 *The Braes of Derwent*, one of the five D49s that would be fitted with steel axleboxes and B1 bogie to improve the riding in 1951, but seen here in 1949 with standard LNER tender, renumbered but before BR emblem on tender. (Real Photographs/MLS Collection)

Below left: D49/2 62770 *The Puckeridge* on the turntable at Scarborough shed, 1950. (MLS Collection)

Below right: **The last days** – D49/2 62753 *The Belvoir* at Starbeck, c1957. (MLS Collection)

Operation

The initial allocation of the twenty-eight engines built in 1927 and 1928 was:

Haymarket:	2
Perth:	2 (incl Lentz oscillating cam poppet valve 329)
St Margaret's:	5
Dundee:	6
Neville Hill:	6 (incl Lentz oscillating cam poppet valve 318, 320 and 322)
York:	7 (incl Lentz oscillating cam poppet valve 327 and 335 and rotary cam poppet valve 336 and 352)

The pioneer 234, allocated to York, spent three weeks on trial on the former Great Central main line, initially between Marylebone and Leicester and later between Manchester and Sheffield. However, no allocation of the class was ever made to that route and the D49s spent their entire careers in the North East or Scotland.

Most of the logs of D49s are on comparatively short runs in the Edinburgh–Glasgow or Leeds–York–Scarborough routes, but there is one well-known extraordinary effort by a 'Shire' on a heavy East Coast express driven by the famous Haymarket driver, McKillop (alias 'Toram Beg'). The circumstances on which such a heavy train was hauled by a single D49 are unknown, as is also the date, though I sense it was no later than 1930. It is probably the nearest that any timer got to experiencing the maximum effort that a D49 could produce.

Newcastle–Edinburgh
249 *Aberdeenshire*
13 chs, 435 tons
1.20pm King's Cross–Edinburgh

Miles	Location	Times	Speed		Gradients
0	Newcastle	00.00		T	
1.7	Heaton	04.26			1/200 R
5	Forest Hill	10.03	37		1/200 R
9.9	Cramlington	16.22	53/48		L, 1/224 R
13.9	Stannington	20.33	66		1/217 F
16.6	Morpeth	23.35	35*	1 L	
18.5	Pegswood	26.05	56		L

Below left: A Scottish D49, 236 *Lancashire*, on a heavy East Coast main line express at an unidentified location, c1930. (Rail Archive Stephenson/John Scott-Morgan Collections)

Below right: 253 *Oxfordshire* departing from Sheffield Victoria with an express for Leeds and Newcastle, c1932. (MLS Collection)

		Newcastle–Edinburgh			
		249 *Aberdeenshire*			
		13 chs, 435 tons			
		1.20pm King's Cross–Edinburgh			
Miles	Location	Times	Speed		Gradients
23.2	Widdrington	30.43	64½/58½		1/208 F, 1/275 R
28.5	Acklington	35.49	61½/67½		1/320 F, undulating, 1/330 F
31.9	Warkworth	39.00	62		undulating, 1/220 F
<u>34.8</u>	<u>Alnmouth</u>	<u>43.10</u>		2¼ L	
0		00.00		2 L	
2.7	Longhoughton	05.59	34		1/170 R
4.6	Little Mill	09.17	36		1/170 R
8.2	Christon Bank	13.27	65		1/150 F
11.2	Chathill	16.09	69		1/573 F
16.8	Belford	21.19	63½	2¾ L	L
23.8	Beal	27.19	76½		1/208 F, 1/300 F
28.7	Scremerston	31.43	58½		1/190 R
30.9	Tweedmouth	34.15	54 eased		1/245 F
<u>32.1</u>	<u>Berwick-on-Tweed</u>	<u>36.22</u>		2¼ L	
0		00.00		2 L	
1.1	Marshall Meadows	03.30	30	2½ L	1/190 R
5.6	Burnmouth	11.18	37½		1/190 R
11.2	Reston Junction	17.24	62½	6 L	1/250 F, 1/200 R
16.3	Grantshouse	23.32	46½	7 L	1/200 R
21	Cockburnspath	28.07	75		1/96 F
24	Innerwick	-	80½		1/210 F
26.2	Oxwell Mains Sdgs	32.14			L
<u>28.4</u>	<u>Dunbar</u>	<u>35.55</u>		5 L	(conditional stop)
0		00.00			
5.7	East Linton	09.02	50½		1/360 R, 1/300 R
8.3	East Fortune	11.45	64½		1/250 F
<u>11.3</u>	<u>Drem</u>	<u>15.32</u>		8½ L	(conditional stop)
0		00.00			
4.6	Longniddry	07.54	60	11½ L	1/300 R, 1/300 F
8.3	Prestonpans	11.45	57/62½		undulating
11.3	Inveresk	14.34	66½		1/300 F
14.8	Portobello	19.12		11¾ L	
<u>17.8</u>	<u>Edinburgh Waverley</u>	<u>24.37</u>		<u>11 L</u>	

The schedule quoted above assumed passing both Dunbar and Drem at speed. Without the conditional stops, the excess over schedule would have been limited to 2-3 minutes at most. Driver McKillop certainly worked 249 very hard on the banks and in view of the poor riding reputation of the D49s, McKiilop is to be commended for running hard down from Grantshouse to Cockburnspath.

The eight 'Shires' built in 1929, 2753–2760, all went to Scotland, one to St Margaret's and the other seven to Eastfield. In May 1932 with the appearance of the D49/2 'Hunts', Neville Hill's Lentz engines were reallocated to Hull Botanic Gardens as was 327 from York. 335 followed a month later. Of the first batch of new 'Hunts, 201, 211, 220, 273 and 282 went to Leeds and 232, 235, 247, 255 and 269 to York. 282, the locomotive with continuous cam to operate its valves, went to Hull in 1933 to operate its turns to Sheffield. In 1933 the next batch of 'Hunts' also went to the north east, 283, 288, 297 and 298 to York and 292 to Hull. Later that year 283 and 297 were transferred to Leeds Heaton depot for working 'The Queen of Scots' Pullman train between Newcastle and Leeds. Neville Hill engines also worked 'The North Briton' express between Leeds and Newcastle.

The *Railway Magazine* of April 1933 reported two interesting snippets of both classes of D49 over the Darlington–York racing stretch with heavy loads for a 4-4-0. They were probably working Newcastle–Leeds expresses that would go on to Sheffield, Banbury and the Great Western, although it is known that D49s did have a turn or two between York and Grantham with East Coast expresses for a short time in the early 1930s. Both runs were with pioneer engines of their sub-classes, Leeds's D49/2, 201 and York's D49/1, 234.

D49/2 211 *The York and Ainsty* preparing to depart from York with a northbound express, c1933. (Real Photographs/MLS Collection)

York's D49/2, 288 *The Percy*, awaits the arrival of 'The Scarborough Flyer' from King's Cross, ready to take over from its Gresley Pacific for the run to Scarborough, c1934. (MLS Collection)

		Darlington–York, c1932				
		201 *The Bramham Moor*		**234** *Yorkshire*		
		12chs, 355/370 tons		15 chs, 452/480 tons		
Miles	Location	Times	Speed	Times	Speed	Gradients
0	Darlington	00.00		00.00		
2.6	Croft Spa	05.15		05.22	49½	1/438 F
5.2	Eryholme	08.14	52	08.33	47	1/391 R
10.4	Danby Wiske	13.27	pws	14.12	54	1/650 F
14.1	Northallerton	17.57		18.15	55	L
17.5	Otterington	21.39	67	21.47	58	1/629 F
21.9	Thirsk	25.31	70	25.55	64½	L
26.1	Sessay	29.11	69	30.06	61	L
28	Pilmoor	30.50	72	32.00	60	1/739 F
30.7	Raskelf	33.07	73½	34.35	62	1/741 F
32.9	Alne	34.55	74	36.35	64½	L
38.6	Beningbrough	39.42	72	42.00	62½	L
42.5	Poppleton Junction	43.00	72	45.47	62	L
44.1	York	46.05	(44½ net)	48.41	¾ L	

2753 *Cheshire* with the Up 'Queen of Scots' Pullman train arriving at Edinburgh from Glasgow, 19 June 1936. (MLS Collection)

The allocation of the final twenty-five 'Hunts' built in 1934/5 was:

Neville Hill:	9
Gateshead:	4
York:	1
Botanic Gardens:	3
Scarborough:	6
Bridlington:	2

The Scottish allocation had little disturbance in the 1930s. The Lentz oscillating cam valve engines were having problems and Scotland's only example, 329, came south to York in 1937 for special observation before the rebuilding of all six in 1938 as D49/1s. 245 *Lincolnshire* from Leeds was sent to Scotland as a replacement and worked on loan from Eastfield for a few months. This engine had spent six months on loan to King's Cross earlier in the autumn/winter of 1928/9 and was the most widely travelled D49. 307 and 310 of Dundee were loaned to Aberdeen for eighteen months in 1932/3, returning to Dundee. The only other significant move in Scotland before the war was the transfer of four D49s, 2754–2757, to Carlisle to replace Reid Atlantics, one in 1933, another in 1936 and the last two in 1937. The Scottish allocation in 1939 was:

Carlisle:	4
St Margaret's:	5
Haymarket:	2
Eastfield:	4
Dundee:	5
Perth:	2
Thornton Junction:	1

The archives of the Rail Performance Society have a large number of logs recorded in the mid-late 1930s with D49s mainly in the north east, especially between Leeds and York, Selby and Hull, York and Scarborough. In Scotland they mainly feature on semi-fast services between Edinburgh and Glasgow. In 1935 D.S. Barrie recorded a typical Edinburgh–Glasgow run:

Haymarket–Glasgow Queen Street, 21.5.1935				
309 *Banffshire* D49/1				
6chs, 210/220 tons				
Miles	Location	Times	Speed	Gradients
0	Haymarket	00.00		T
2.3	Saughton Junction	04.33	45	1/960 R
7.3	Bathgate	10.55	54/ pws 30*	1 L
11.3	Winchburgh Jcn	16.14	45	1¼ L
13.3	Philipstoun	18.47	58	1/882 F
16.4	Linlithgow	22.35		1 L
0		00.00		1 L
2.2	Manuel	03.58	58	1/882 R
4.7	Polmont	06.32	60	1/882 R
7.9	Falkirk High	10.10		1¼ E
0		00.00		T
4.5	Greenhill Junction	07.06	57	T L
10.3	Croy	12.53	69/ pws 15*	1/900 F, L
15.5	Lenzie	19.47	46/59	2¾ L
19.9	Cowlairs	24.32		1/41 F
21.8	Glasgow Queen St	30.48		1¾ L

J. Wedgwood recorded a number of runs in the Selby/Doncaster–Hull area in 1935/6. Three runs between Selby and Hull (30.8 miles) were completed by 253 *Oxfordshire* and 228 tons in a net 37 minutes against a 39 minute schedule, and against a faster 36 minute schedule, 205 *The Albrighton* in 35¼ minutes net with 254 tons and 322 *Huntingdonshire* in 35½ minutes net with 236 tons. Between Broomfleet and Brough on the level, maximum speeds were 60mph (253), 65 (322) and an exceptional 72 (205). A couple of runs on the 2.55pm Doncaster–Hull in 1936 were logged by Wedgwood. 214 *The Atherstone* and 203 tons ran the 17.2 miles from Doncaster to Goole in 21½ minutes, maximum speed 55mph and the 23.7 miles onto Hull in 30 minutes 5 seconds, maximum 63mph. 318 *Cambridgeshire* with a heavy 385 ton load ran the 40.9 miles non-stop in 54 minutes (49 net) but both

270 *Argyllshire* departing from Edinburgh through Princes Street Gardens with an express for Dundee and Aberdeen, 18 June 1936. (MLS Collection)

306 *Roxburghshire* with an Aberdeen–Edinburgh express passing the Princes Street Gardens, 4 August 1938. (L. Hanson/MLS Collection)

264 *Stirlingshire* double-heads a D11/2 'Director' across the Forth Bridge with an Aberdeen–Edinburgh express, c1938. (Photomatic/MLS Collection)

injectors failed at Staddlethorpe, the D49/1 just managing to finish the journey before being failed. In the reverse direction, 377 *The Tynedale* and just four coaches on the 4.30pm Hull completed the non-stop run in 47 minutes 10 seconds (schedule 50 minutes) and 205 *The Albrighton* with 9 coaches ran the 23.7 miles from Hull to Goole in 29 minutes (28½ net) against the 32 minute schedule, maximum speed 68mph. On the Scarborough run, 258 *The Cattisbrook* with seven coaches (180 tons) ran the 21 miles from Scarborough to Malton in 22¼ minutes (schedule 25), maximum speed 68mph, and the 21 miles on to York in 25¾ minutes (schedule 27), maximum speed 63mph.

Messrs. Manners, Gelder and Wedgwood submitted to the RPS a vast number of logs of all types of LNER power (V2, C1, B16, D20, C7, A3 & A4) between York and Leeds in both directions. Between 1936 and 1941 in excess of 500 logs with D49s were compiled with a range of times, mostly around 30–31 minutes net for the 25.6 miles which includes some five miles of climbing to Garforth, mainly at 1 in 160/162 through Cross Gates eastbound and 1 in 145/133 through Micklefield westbound. The ten miles between Church Fenton and York is level. A sample of some of the best runs I can find eastbound from Leeds are outlined below:

Loco	318	235	226	374	375
Name	*Cambridgeshire*	*The Bedale*	*The Bilsdale*	*The Sinnington*	*The South Durham*
Type	D49/1	D49/2	D49/2	D49/2	D49/2
Load	183 tons	206 tons	198 tons	231 tons	231 tons
Act time	29.43	28.43	29.37	28.19	28.47
Max speed (mph)	71	80	66 ½	77	72
Location	Copmanthorpe	Church Fenton	Church Fenton	Church Fenton	Church Fenton

Schedules ranged from 30–32 minutes, decelerated to 35 minutes in the war years. One detailed log is tabled below, typical of the better runs:

		Leeds–York, 25.9.1938		
		368 *The Puckeridge* D49/2		
		7 chs, 236 tons		
Miles	Location	Times	Speed	Gradients
0	Leeds	00.00		
0.8	Marsh Lane	02.23		L
2.7	Osmanthorpe	05.35	35	1/153 R
4.4	Cross Gates	08.15	39	1/160 R
7.3	Garforth	12.10	44½	L
9.7	Micklefield	14.37	57	1/150 F
14.8	Church Fenton	18.35	77	1/133 F, 1/145 F
16.3	Ulleskelf	20.08	73½	L
17.9	Bolton Percy	21.11	68½	L
21.8	Copmanthorpe	24.30	70½	L
23.6	Chaloners Whin Jcn	26.05	68/sigs	L
25.6	York	30.10	(28¾ net)	

The fastest westbound was a run in 27 minutes net but with a featherweight load of four coaches only:

		York–Leeds, c1937		
		357 *The Fernie* D49/2		
		4 chs, 106/110 tons		
		7pm Leeds–York		
Miles	Location	Times	Speed	Gradients
0	York	00.00		
2	Chaloner's Whin	03.00		L
3.8	Copmanthorpe	05.00	60	L
7.7	Bolton Percy	08.25	70	L
9.1	Ulleskelf	09.30	74	L
10.8	Church Fentton	11.00	68	L
15.9	Micklefield	15.35	60	1/145 R, 1/133 R
18.3	Garforth	17.55	62	1/150 R, L
21.2	Cross Gates	20.20	75	L, 1/160 F
22.9	Osmanthorpe	24.18	sigs 20*	1/162 F
24.8	Marsh Lane	26.35	sigs 10*	
25.6	Leeds	29.50	(27 net)	

269 *The Cleveland* arriving at Sheffield Victoria with a Newcastle to West of England express consisting of GWR rolling stock, 23 July 1936. (MLS Collection)

A pair of D49/2s, 365 *The Morpeth* and 376 *The Staintondale* leaving York bound for Leeds, c1937. (MLS Collection)

In 1939, a major redistribution of locomotives in the north east was made to reduce the number of different classes at depots, the impact on the D49 class being that the D49/2s were concentrated at Neville Hill Leeds (15), York (10), Gateshead (8) and Scarborough (5) with just four at Hull. All eleven D49/1s in the north east were allocated to Hull Botanic Gardens.

Loads increased in the war years and additional stops were included, and frequent signal checks meant that the Leeds–York wartime schedules of 35 or 40 minutes were usually exceeded. On 11 May 1940, 370 *The Rufford* and 375 *The South Durham* double-headed a 13 coach 415 ton 12noon Harrogate–Darlington, running the 11½ miles to Ripon in 19 minutes as scheduled, but net 13¼ minutes and the 27¾ miles on to Darlington in 33¾ minutes. On 25 May 1940, 374 *The Sinnington* with 13 coaches, 365 tons, on the 2.51pm from Durham covered the 22 miles south to Darlington in 29 minutes 35 seconds (schedule 32 minutes) with speed mainly in the 50s, just touching 60 between Aycliffe and Springfield down the 1 in 200.

Little change took place in Scotland during the war years until 1943, when Haymarket's allocation

375 *The South Durham* at Greatham with a Leeds–West Hartlepool train, c1938. (Real photographs/MLS Collection)

With war looking, the D49s take on freight duties – 362 *The Goathland* with a heavy partially fitted freight near Alnmouth, c1939. (MLS Collection)

was boosted to twelve to replace NE and NB Atlantics, so that its full allocation of D49/1s was:

Carlisle:	4
Haymarket:	12
Dundee:	3
Perth:	3
Thornton Junction:	3

Despite the fact that the D49/1s put in some hard work in Scotland, they were not popular because they were rough-riding and draughty – their replacement of the more comfortable Reid Atlantics did not go down well with the crews. The criticisms were met with the transfer of more Gresley Pacifics to Haymarket and V2 2-6-2s to St Margaret's and the Edinburgh D49s were mainly seen only on express turns on the Edinburgh–Glasgow turns, relegated to slower semi-fast and stopping trains elsewhere.

During the war, the D49s were utilised for a considerable amount of goods work despite their unsuitability as they could only match the loads of the J35 0-6-0s. The main line work on passenger trains was limited to stopping services and they could not be used on branch services because of the high axleloads and route availability. The north eastern allocation during the war remained unchanged from the big readjustments in 1939. 253 and 320 had gone from Hull to Scotland in 1943 in exchange for two NE C7s. In the north east the D49/2s struggled with the increased wartime loads. The Leeds–Newcastle trains needed a pair of D49s, even on the slower schedules. The Leeds–York schedules were decelerated from 30 to 35 minutes for the 25½ miles, but the D49s regularly were required to haul 350 or even 400 tons on this stretch, usually losing a minute or two if loaded to this level.

In the immediate aftermath of the war, with the influx of

Thompson B1s into the north east, the first D49 allocations were made to Harrogate Starbeck, the renumbered 2726, 2752, 2753, 2762 and 2773. They were joined by the Thompson rebuild 'D', 2768, which had worked initially at Haymarket on D11 turns. 2768 spent the rest of its short career there. At nationalisation, the north eastern D49s were allocated to:

Leeds Neville Hill:	10 (D49/2)
York:	7 (D49/2)
Gateshead:	12 (D49/2)
Starbeck:	5 (D49/2)
Hull Botanic Gdns:	7 (D49/2), 9 (D49/1)

An attempt to cure the rough-riding of the D49s in 1951 with B1 type bogies was made with five D49/2s, and it was decided to test these in both the north east and Scotland. Two of the modified D49s were sent, 62743 from Gateshead to Haymarket and 62744 from York to Dundee. In exchange two Scottish D49/1s, 62702 and 62717, were sent to the North Eastern Region. Only one other inter-regional exchange took place, 62747 going to Carlisle, and its 62732 to Darlington.

The allocation of the entire class of D49/1s at the beginning of 1950 was:

Haymarket:	62702*, 62705, 62706, 62709, 62719, 62733
St Margaret's:	62711, 62712, 62715, 62721
Thornton Junction:	62704, 62708, 62716, 62717*, 62729
Dundee:	62713, 62718, 62728
Perth:	62714, 62725
Carlisle Canal:	62730–62732*, 62734, 62735
Hull Botanic Gdns:	62700, 62703, 62710, 62720, 62722–62724
Bridlington:	62701, 62707

(* inter-regional subsequent adjustments as noted in previous paragraph.)

The allocation of the D49/2 'Hunts' was:

York:	62726, 62727, 62736, 62740, 62742, 62744*, 62745, 62759 – 62761
Hull Botanic Gdns:	62737, 62741, 62743*, 62754, 62757, 62767
Leeds Neville Hill:	62739, 62746, 62748, 62756, 62775
Starbeck:	62738, 62749, 62752, 62753, 62755, 62758, 62762, 62763, 62765, 62768, 62772, 62773
Blaydon:	62747*, 62771
Bridlington:	62750, 62766
Scarborough:	62751, 62764, 62769, 62770
Pickering:	62774

(* inter-regional subsequent adjustments as noted in previous paragraph.)

62731 *Selkirkshire* shortly after its transfer from Carlisle Canal to York via a month's stay at Starbeck, entering Scarborough station with a train from Leeds, summer 1951. (MLS Collection)

62725 *Inverness-shire* arriving at Perth with a train from Dundee, 30 May 1953. (MLS Collection)

62730 *Berkshire* transferred with 62731 from Carlisle Canal to York via Starbeck in the autumn of 1950, leaving York with a Scarborough train, c1954. (MLS Collection)

The pioneer D49/2 built in 1929 as 62726 *Leicestershire* with rotary cam poppet valves and renamed *The Meynell* in June 1932, at Church Fenton with a York–Leeds stopping train, 17 May 1952. (C.H.A. Townley/ MLS Collection)

York's 62745 *The Hurworth* at Malton, c1952. (MLS Collection)

The pioneer D49/2 again, York's 62726 *The Meynell* at Escrick with the 4.40pm York–Doncaster stopping train, 9 May 1953. (J.D. Darby/MLS Collection)

Neville Hill's 62748 *The Southwold* entering York with a stopping train from Leeds, c1955. (MLS Collection)

Researching the Rail Performance Society archives for post-war D49 logs produced many but mainly of a fairly pedestrian nature. Few records seem to have been taken of performance by the Scottish Region based D49s. O.S. Nock described a footplate run on the 8.45am Edinburgh–Perth with 62702 *Oxfordshire* in October 1951 with a load of six coaches, 175 tons. He described the engine climbing 'beautifully' with the use of 45 per cent cut-off but the riding was poor, the engine 'snaking'. Hugh Gould timed 62708 *Argyllshire* with seven coaches, 229 tons, on the 12.51pm Glasgow Queen St–Kirkcaldy (a Fife Coast train). It topped Cowlairs Bank at 31mph in just over 6 minutes as scheduled and with speed only in the mid-50s had dropped a minute by Greenhill Junction, but 62mph after saw the train into Falkirk in 30¼ minutes from Glasgow, 2 minutes early. A six-minute run to Polmont gained another minute and a twenty minute run onto Dalmeny cut the schedule by another two minutes. Gould left the train at Inverkeithing, reached 2½ minute early, the result of slack schedules rather than exceptional performance, with no speed over 59mph after Falkirk. E.R. Davies timed 62725 *Inverness-shire* on the 1.39pm Larbert–Edinburgh Princes Street (a train from Perth) which attained 62mph at Winchburgh Junction after the Falkirk stop, and 60 at Gogar after a p-way slack to 32mph, arriving at Princes Street a minute early, but with only four coaches.

In England many more logs were recorded, but mainly on semi-fast or stopping services. D.S. Barrie logged 336 *The Quorn* in October 1944 with a 10-coach load of 320 tons covering the 18 miles from Seamer to Malton in 3 seconds under the scheduled 23 minutes (max 57mph) and taking just over 33 minutes for the 21.1 miles to York, 56 maximum speed, dropping 3 minutes on the schedule. P.W.B. Semmens timed Harrogate–Stockton and York–Harrogate trains with 62773 and 62763 respectively, but only on 4-5 coach loads and speed maxima in the mid-60s on the longer sections between stops. Derek Twibell timed 62700 *Yorkshire* in July 1951 on a York–Leeds train, load 9 coaches, 270 tons, completing the 21 miles to the first stop at Cross Gates in 27¾ minutes, top speed 65½mph on the level at Church Fenton before the climb to Garforth. A couple of runs with Carlisle Canal's 2732E *Dumfries-shire* were timed between Carlisle and Haltwhistle, but with just six coaches and several stops, timekeeping was maintained with nothing over 55mph.

Noel Proudlock of the Rail Performance Society wrote an article in their magazine *Milepost* on the D49/2 'Hunt' class, particularly pointing out the weakness of the limited fixed five point cut-off arrangement of the poppet valves. He suggested that often the inability to make finer adjustments led to loss of time, or excessive fuel consumption with working harder than was necessary. He referred to a number of calculations made around 1955 of maximum power output, with single D49/2 drawbar horsepower ranging from 600–800 maximum, the latter at a sustained 44mph (62727 with 204 tons between Ripon and Wormald Junction). Another effort with 62740 on 9 coaches, 310 tons got stopped on the 1 in 100 on the Ripon–Wormald Junction section and could produce no more than 625 hp as it struggled to restart, slipping badly. Pairs of D49/2s were used on the 325 ton 4.22pm Scarborough–Bradford and three consecutive runs with 62762 & 62642, 62762 & 62727 and 62762 & 62740 produced around 1,200edhp on the 1 in 133 climb from Church Fenton through Micklefield at a sustained 44mph. Proudlock had a generally poor opinion of the D49s post-war, with their poor riding and propensity to slip, recording many cases of lost time and the highest speed he experienced with a D49 was 72mph on the descent from Micklefield to Church Fenton. He mentioned a run with 62748 which, admittedly in wet weather, slipped and slithered with its 234 ton load on the 1 in 94 all the way from Arthington to Horsforth. He expressed the view that the Thompson B1 4-6-0s could match the performance of the pair of D49s on the Leeds–Bradford train, and the increase in numbers of the B1s in the 1950s replaced the use of D49s on anything but lightweight stopping services.

As the 1950s progressed, the Scottish D49/1s rarely strayed from their home depots, although the Perth pair moved to Stirling. In the north east the B1s had taken over much of their former work and by 1955 they were relegated to slow stopping passenger trains or even freights for which they were hardly suitable. In the last year, 1956, in which the class was still intact – apart from the 2-cylinder rebuilt 62768 – their allocation was (comparison with 1950 in brackets):

D49/1s:
Haymarket: 62705, 62706,
               62709, 62719,
               62733 (-1)
St Margaret's: 62711, 62712,
               62715, 62718,
               62721 (+1)
Thornton
 Junction: 62704, 62708,
               62713,
               62716, 62729
               (No change)
Dundee: 62728 (-1)
Stirling: 62714, 62725 (+2,
               Perth -2)
Carlisle Canal: 62732, 62734 (-3)
York: 62702, 62730,
               62731 (+3)
Hull Botanic
 Gdns: 62700, 62707,
               62710, 62717,
               62720, 62722 –
               62724 (+1)

Bridlington: 62701, 62703 (No change)
Scarborough: 62735 (-3)

Allocation of D49/2s in 1956:

Haymarket: 62743 (+1)
Dundee: 62744 (+1)
York: 62745, 62747, 62760, 62771 (-6)
Hull Botanic Gdns: 62737, 62741, 62754, 62757, 62766, 62767 (No change)
Leeds Neville Hill: 62740, 62742, 62748, 62749, 62755, 62761, 62772, 62775 (+3)
Starbeck: 62727, 62736, 62738, 62746, 62749, 62752, 62753, 62758, 62759, 62762, 62763, 62765, 62768, 62773, 62774 (+3)
Bridlington: 62750 (-1)
Scarborough: 62726, 62739, 62751, 62756, 62764, 62769, 62770 (+3, Pickering -1)

Withdrawals in both the Scottish and North Eastern Regions began in 1957, just two D49/1s, 62713 from Thornton Junction and 62724 from Hull Botanic Gardens. Five D49/2s were withdrawn that year, 62726 from Scarborough, 62748 and 62761 from Neville Hill, 62757 from Botanic Gardens and 62758 from Starbeck. Ten D49/1s and twenty D49/2s went in 1958; ten D49/1s and six D49/2s in 1959; three D49/1s and five D49/2s in 1960; and finally nine D49/1s and five D49/2s in 1961. The last survivors in traffic were both Scottish D49/1s - 62711 from St Margaret's and 62729 from Thornton Junction. The last to be officially withdrawn was 62712 in July but from March it had been acting as a stationary boiler in Edinburgh which it continued to do until January 1962, then being in store pending preservation (see chapter 7).

Botanic Gardens' 62717 *Banffshire* performs station pilot duties at Hull Paragon station, c1956. (MLS Collection)

St Margaret's
62718 *Kinross-shire* pilots a Gateshead A1 with a heavy East Coast express on Grantshouse Bank, c1958. (MLS Collection)

Starbeck's 62759 *The Craven* at Leeds on the 5.30pm to Harrogate, c1956. (B.K.B. Green/MLS Collection)

Starbeck's 62763 *The Fitzwilliam* on the 10.20am Leeds–York stopping train, 6 August 1956. (A.C. Gilbert/MLS Collection)

York's 62760 *The Cotswold* on the Harrogate portion of the 1.18pm King's Cross which it worked from Leeds, near Horsforth, 26 May 1958. 62760 moved to Bridlington a month later. (B.K.B. Green/MLS Collection)

Chapter 6
PERSONAL EXPERIENCES

My first sighting of an LNER 4-4-0 was in August 1950 when I accompanied a 12 year old friend on his first holiday without his parents to stay with an aunt in the Doncaster area. I think the idea was to accompany him on his visits to relatives still living in Yorkshire combined with a few trainspotting trips. My memory was that the latter outnumbered the dutiful appearances to various grandparents, aunts and uncles. Most days we joined other trainspotters in the cattle dock grids at Doncaster station but we took a few bus trips – we couldn't afford trains unfortunately – to places like Sheffield and York. On the trip to the latter, I vaguely recollect seeing 62751 *The Albrighton* high above as we walked along a road just north of the station. I mentioned in my previous LNER 4-4-0 book the sighting of a couple of 'Directors' at Doncaster on locals from Sheffield. A holiday at Whitby in 1955 included a day in York when I photographed a couple of D49/2s on trains for Scarborough, although my own travels included only Raven A8 tanks, a B16/3 on *'The Scarborough Flyer'* and ubiquitous B1s. I must have seen a couple of D20s (former NER 'Rs') as numbers were underlined in my Ian Allan ABC book, though I fail to recall where – I think York itself was most likely.

A memorable experience of these engines occurred in 1957 and even more memorably in 1958. In September 1957 I spent a week at the Methodist Guild Holiday Home in Dunoon, 'Dhalling Moor', travelling up from London behind 46244 *King George VI* on *'The Royal Scot'* and returning the following Saturday on *'The Midday Scot'* behind a couple of 'Princess Royals', 46203 from Glasgow to Crewe and 46209 on to Euston. The weather had been poor that year – three days in a Scotch mist with the Clyde shipping siren booming mournful warnings reverberating in the clammy air, while we students looked for opportunities to hike in the glens. I escaped one dismal day when the hike was cancelled and chose to go to Edinburgh where, surprisingly, I found sunshine. The idea was to photograph trains at Edinburgh Waverley station and from the Princes Street Gardens. I took the ferry from Dunoon to Craigendoran, picked up a commuter train to Glasgow Queen Street behind V1 2-6-2T 67601, and to my joy found my onward journey to Edinburgh on *'The Queen of Scots'* Pullman headed by that rarity to southern enthusiasts, 60004 *William Whitelaw*. I made my way out to the Princes Street Gardens and spent a couple of hours in that location beloved of so many railway photographers whose work has already adorned this book. First was V2 60822 on the 1.00pm to Aberdeen, then a V1 tank on an excursion, a light engine, named B1 *Strang Steel*, a J37 on a local freight, another V1 and finally a D49/1 62704 *Stirlingshire* before making my way back to Waverley. There I found a D30 'Scott' 62421 *Laird o' Monkbarns* performing station duties at the south end and 62714 *Perthshire* shunting coaching stock at the Haymarket end. I finally spied D49/2 *The Cleveland*, one of the two D49/2s in Scotland, waiting to depart on a stopping train to Dundee, before wending my way back to Glasgow on a Swindon built intercity DMU.

I decided to give it another go in July 1958 and booked a fortnight at 'Dhalling Moor' and hit a rare heatwave for most of the time! For variety I travelled north via the East Coast on *'The Elizabethan'* with my favourite A4, 60033 *Seagull*. I took the middle Saturday off from the student hikes as no organised trips were arranged for the holiday change-over day. I was tempted to try to replicate my 1957 successful Saturday visit and caught *'The Queen of Scots'* once more, this time with Haymarket A1 60159 *Bonnie Dundee*. Before making my intended

Right: **62743** *The Cleveland* waiting to depart from Edinburgh Waverley with an afternoon stopping train for Dundee, 19 September 1957. 62743 was one of the two D49/2s sent to Scotland in the early 1950s in exchange for two D49/1s after five had been modified to improve their riding. (David Maidment)

Below: **62704** *Stirlingshire* arriving at Edinburgh Waverley from Dundee, passing through Princes Street Gardens, 19 September 1957. (David Maidment)

Below right: **D30 62421** *Laird o' Monkbarns,* one of the two last survivors of the class, on station pilot duties at the south end of Edinburgh Waverley station, 19 September 1957. (David Maidment)

way to the Princes Street Gardens, I looked round Waverley station and saw to my astonishment a D34 'Glen' bearing express code headlamps standing facing north at the head of a rake of non-corridor coaches and having ascertained that it was on a train whose first stop was Inverkeithing, I rushed to the ticket office and bought a cheap-day return. I scrambled back and joined the front coach of the train just in time for it was the 11.12am SO relief train to Thornton Junction. The locomotive was 62487 *Glen Arklet* of St Margaret's shed and we made our somewhat pedestrian way past Dalmeny over the Forth Bridge, to me the highlight of the trip, the carriage door window dropped open as I was nearly knocked breathless as we rumbled over the bridge. I had time at Inverkeithing to study the 'Glen' and take a photo as we were a minute or so early, and then, after its departure, wondered what service I had to return, as I'd had no time to consult the timetable earlier. An A3, 60035 *Windsor Lad*, passed through on an Aberdeen–Edinburgh express followed by V2 60916 on a freight and then, to my surprise another 'Glen' rushed into view, 62488 *Glen Alladale* on the main 11.30am Edinburgh Waverley to Thornton Junction.

No sooner had this departed than a Dundee–Edinburgh stopping train sidled noiselessly into the station with D49/1 62708 *Argyllshire* at its head and I enjoyed the return trip over the Forth Bridge behind my second LNER 4-4-0 of the day. On arrival back at Waverley and a snack in the refreshment room, I made my way to the Princes Street Gardens and took a succession of photos – five Pacifics inside an hour – 60535, 60097, 60011, 60528 and 60536, plus a V2, V1 tank and D49/1 62721 *Warwickshire*. Back at the station, I saw other timetabled northbound locals and decided to give it another go, so bought my second day return to Inverkeithing. I was a little disappointed to find my train this time was hauled by a green V2, 60969, but as we left Waverley, I spied a D11/2 'Director' backing down onto a train. I should have waited. When my V2 train stopped at Dalmeny, I wondered if the D11's train would too, so I gambled and got out, risking being stranded as the D11 passed by. But no, my luck held, and a few minutes later 62677 *Edie Ochiltree* slid into the platform and stopped. I boarded and once more rejoiced in my luck as I gulped in the biting air as we passed over the bridge. On arrival at Inverkeithing I took a photo and ascertained that I had travelled on the 2.55pm SO Edinburgh Waverley–Dundee stopping train. In direct contrast, my return local turned up behind a spotless Haymarket A1, 60162 *St Johnstoun* and I arrived back in Edinburgh in good time to catch the 5.15pm Edinburgh–Glasgow express which I had caught the previous Saturday behind Haymarket A3 60087 *Blenheim*.

62714 *Perthshire* shunting coaches at the north end of Edinburgh Waverley, 19 September 1957. (David Maidment)

62487 *Glen Arklet* at Inverkeithing with the Saturdays Only 11.12am relief train to Thornton Junction, 12 July 1958. (David Maidment)

Anyway, I was in luck again as spotless A3 60037 *Hyperion* stood at the head of the Glasgow express. I rounded off the day on an evening Glasgow–Helensburgh train, first stop Dumbarton, behind V1 tank, 67655 and with the ferry to finish, arrived just in time to get a late dinner with newly arrived guests.

When the weather broke in the middle of the second week, I could not resist anther visit to Edinburgh. However, I was faced with poor visibility and steady rain which deterred me from more photographic work in Princes Street Gardens. I'd been told of a group of stored D11s out at Longniddry and caught a local Dunbar train and trudged to the storage sidings at Longniddry in a downpour where I found Scottish 'Directors' 62682, 62685 and 62693 plus a couple of C16 4-4-2 tanks. After a few shots, I hastily returned to Waverley and caught the 5.15 back to Glasgow, with another Haymarket A3, *Spion Kop*.

The D49/1
62708 *Argyllshire* that took me from Inverkeithing back to Edinburgh Waverley in July 1958, seen here at St Margaret's, June 1950. (J.Davenport/MLS Collection)

Scottish 'Director' D11/2 62677 *Edie Ochiltree*, one of three not in store that summer (the others were 62678 and 62679) on the 2.55pm Saturday Only Edinburgh Waverley–Dundee stopping train, 12 July 1958. (David Maidment)

Chapter 7
PRESERVED LOCOMOTIVES

North Eastern Railway 'M', 1621

1621 was the second of Wilson Worsdell's class 'M' locomotives built at Gateshead in 1893 at a cost of £3,110. It and its sister engine 1620 both distinguished themselves in the 1895 'railway races' to the north, 1621 covering the York–Newcastle stretch in 78½ minutes at an average speed of 61.5mph. 1620 was used on the Newcastle–Edinburgh section and covered the 124½ miles in 113 minutes at an average speed of 66mph. At the Grouping it was classified as LNER D17/1. It had been the intention to preserve 1620 when it was withdrawn in 1934, but that plan failed and it was broken up. 1621, however, lasted to July 1945 and at its withdrawal, having run 1,543,407 miles in traffic, it was set aside for preservation. It was restored in the

The record breaking NER 'M' 1621, later LNER D17/1, still in NER livery in the 1920s. What does the disc '55' mean? Was it an exhibit in the Darlington Centenary locomotive cavalcade? It does not look in good enough condition for that. (Real photographs/MLS Collection)

1621 as built and as now exhibited at the Shildon site of the National Railway Museum. (MLS Collection)

North Eastern Railway livery and resided for many years in the small railway museum set up by the LNER beside York station. It moved to the National Railway Museum at York in 1975 and is currently displayed at the NRM 'Locomotion' Museum at Shildon.

North British Railway 256 (62469) 'Glen Douglas'

256 *Glen Douglas* was the third of Reid's final mixed traffic locomotives of class 'K', built at Cowlairs in September 1913. The class was associated for many years with operation over the West Highland line and was classed as a D34 at the Grouping and renumbered 9256. At nationalisation, renumbered 62469, it was still based at Eastfield whose diagrams included turns over the West Highland. However, it was included with a number of other D34s in a transfer to the former Great North of Scotland lines as older engines of that company were withdrawn, being allocated to Kittybrewster in May 1953 and Keith in February 1956. It was withdrawn in November 1959 and based at Glasgow's Dawsholm depot, running a number of special railtours in the 1960s. It had been donated to the Glasgow Corporation for inclusion in their Museum of Transport where it was placed after the tours ceased.

256 was involved in *'The Borders Railtour'* in July 1961 after 46247's spectacular climb to Ribblehead, taking over from a couple of B1s at Hawick and running the special to Tweedmouth including the branches from St Boswell's to Greenlaw and Roxburgh Junction to Jedburgh. From Roxburgh Junction through Kelso and Coldstream 256 ran at a steady 36-38mph just about maintaining point to point times without regaining any of the twenty minutes lost on the branch turnrounds. At Tweedmouth it appropriately handed the train over to A1 60143 *Sir Walter Scott*. Another ambitious railtour took place on 1 June 1963, when a full day excursion from Glasgow Queen Street to Mallaig and back was planned with 256 and a J37 pilot to Fort William and back and a pair of J37s on the Mallaig extension.

2469 *Glen Douglas* runs into Glenfarg station with the 4.30pm Perth–Ladybank, 23 July 1949. (J.D. Darby/MLS Collection)

256 *Glen Douglas* on a railtour in the early 1960s at an unknown location. (A.C. Gilbert/MLS Collection)

The train left Queen Street with nine full coaches at 7.50am banked by a North British Type 2 diesel to Cowlairs and ran along the Clyde Coast to Helensburgh Upper, reaching 53mph and 51mph between signal checks. Unfortunately the J37 pilot ran hot after Bridge of Orchy and was removed at Gorton, leaving 256 to continue alone but with a load too great to tackle the steep grades from Rannoch to Corrour, so we (I was a passenger on the train) picnicked in the heather while waiting for a diesel from Fort William to come to our assistance. 256 was failed at Fort William, its brick arch having collapsed. One of the pair of J37s also ran hot en route to Mallaig, so the train returned very late behind diesel traction.

It was later loaned to the Scottish Railway Preservation Society at their Bo'ness Museum but plans to restore it to operation to celebrate the centenary of the West Highland line in 1994 came to naught. It was returned to Glasgow in 2008 and in 2011 it was displayed at the Glasgow Riverside Museum.

Great North of Scotland Railway 49 (62277) 'Gordon Highlander'

Pickersgill designed the first 'V' class 4-4-0 in 1899 and six were constructed by Neilson & Co. A further eight were built at Inverurie Works between 1909 and 1913, and finally six superheated versions, class 'F', were built by the North British Loco Company in 1920 and two at Inverurie Works in 1921. No.49 *Gordon Highlander* was one of the North British superheated 'F' class and was built in October 1920. It was renumbered 6849 at the Grouping, 2277 in 1946 and 62277 at

6849 *Gordon Highlander* as seen at Aberdeen in the 1930s. (J.A.G. Coltas/MLS Collection)

2277 *Gordon Highlander* at Kittybrewster in August 1947. (MLS Collection)

nationalisation in 1948 when it was based at Kittybrewster depot. It was transferred to Keith in June 1951 from where it was withdrawn as the last member of the D40 class in June 1958.

It was restored at Inverurie and painted in the Great North of Scotland green pre-Heywood livery which it never carried. It was based at Glasgow Dawsholm depot whilst working railtours in the early 1960s alongside 256, the Caledonian single 123 and the GWR *City of Truro*. Among its railtours was one in October 1965 from Edinburgh Waverley to Glasgow St Enoch via Bathgate, Airdrie, Lanark and Carstairs. In June 1966 it was installed in the Glasgow Museum of Transport but was then loaned to the Scottish Railway Preservation Society and is currently displayed (in 2023) at their museum at Bo'ness.

62277 *Gordon Highlander* just before withdrawal and preservation, c1957. (MLS Collection)

49 *Gordon Highlander* as preserved and working the Branch Line Society railtour of 16 October 1965. (A.C. Gilbert/ MLS Collection)

LNER D49/1 246 (62712) 'Morayshire'

246 *Morayshire* was built at Darlington in February 1928 and was one of the class that spent its entire career in Scotland. At various times in the 1930s it was based at Dundee, Perth, Haymarket and Hawick. It surprisingly received new frames and cylinders in 1935 and had four boiler changes in its career. At nationalisation in 1948 it was based at Haymarket and was moved to St Margaret's in March, where it stayed ten years before spending a couple of years from February 1958 at Thornton Junction, finishing at Hawick in March 1960, being withdrawn four months later.

It was used then for six months as a stationary boiler for Slateford Laundry near Edinburgh and then stored at Dalry Road while Ian Fraser negotiated for its purchase.

In 1964, it was hauled to Inverurie and restored and held at ICI Ardeer until 1966, then at the Royal Elizabeth Dockyard at Dalmeny from where it was handed over to the Royal Scottish Museum in Edinburgh. In 1974 it was loaned to the Scottish Railway Preservation Society, was restored to operation in steam for the 150th anniversary of the Stockton & Darlington Railway. It was taken out of service in 1981, and dismantled at Falkirk. Further restoration restarted in early 2000 and was completed in 2003. It operated at Bo'ness from 2005 to 2011 when it was withdrawn after its boiler certificate expired. In 2014 it was returned to traffic in BR mixed traffic lined black livery.

Withdrawn again in 2016, it was contracted to the Llangollen Railway for overhaul, but although initial overhaul took place, there was a dispute over fees and the quality of work carried out and 246 left Llangollen in 2020 and returned to Bo'ness to be overhauled to the required standard, estimated to cost £100,000. In 2021, it moved to Locomotive Maintenance Services in Loughborough, and boiler repairs were completed by November. In 2023, £25,000 is needed to complete the overhaul when it is anticipated it will be restored to LNER lined green livery.

Like the other preserved Scottish 4-4-0s, it did some railtour work, but not until the early 1980s. It ran a SRPS railtour in September 1980 and another in April 1981, both Falkirk–Dundee routes. It worked on a number over heritage railways during its operational years including on both the Great Central and Llangollen Railways.

62712 *Morayshire* on an Edinburgh–Carlisle stopping train at Fountainhall Junction, 7 September 1950. (J.D. Darby/MLS Collection)

62712 *Morayshire* in store at Thornton Junction, 6 April 1958. (MLS Collection)

Morayshire as restored in the early 1980s in LNER lined green livery as 246. (A.C. Gilbert/ MLS Collection)

COLOUR SECTION

NER 'R' (LNER D20/1) 1207, built 1907, (later BR 62386) in the Clifton Carriage Sidings at York, 1937. (Colour Rail)

LNER D20/1 62378 (formerly NER 'R' 724) at York with a train from Leeds, 1955. (Colour Rail)

D20/1 62387 at Leeds before departing for York and Alne with the RCTS *Yorkshire Coast Rail Tour*, where it will hand over to D49/1 Selkirkshire, 23 June 1957.
(Transport OnLine Collection)

D20/2 62360 at Northallerton on the SLS/MLS *Northern Dales Rail Tour*, awaiting the attachment of A8 4-6-2T 69855 before setting forth for Hawes and Garsdale, 4 September 1955. (MLS Collection)

D20/1 62387 at Alne with the RCTS *Yorkshire Coast Rail Tour*, 23 June 1957. (Transport OnLine Collection)

D30 62423 *Dugald Dalgetty* on a local service at Humshaugh, between Reedsmouth and Hexham in Northumberland, 1953. (Colour Rail)

D30 62441 *Simon Glover* with a Dundee–Edinburgh stopping train at North Queensferry, c1956. (Colour Rail)

D34 9035 *Glen Gloy* in the LNER black with red lining livery of 1928, at Glasgow Eastfield, August 1939. (Colour Rail)

D34 62478 *Glen Quoich* with a local train for Inverkeithing and Dundee at Haymarket West, 1957. (Colour Rail)

D34 62488 *Glen Aladale* at Whitrope summit with a Galashiels–Hawick stopping train, 25 June 1960. (Colour Rail)

62496 *Glen Loy* leads 62471 *Glen Falloch* at Crianlarich with the 5.45am Glasgow–Fort William express being filmed for the BBC programme *Railway Roundabout* in May 1959. (MLS Collection)

Great North of Scotland D40 62271, a Pickersgill class 'V' built in 1914, at Craigellachie with the 2.55pm to Boat of Garten, 28 May 1955. (Transport OnLine Collection)

62271 moves off the turntable at Boat of Garten, 2 April 1956. (Transport OnLine Collection)

D40 62277 *Gordon Highlander,* of the last series of Pickersill 4-4-0s, class 'F', built by the North British Loco. Co. in 1920, on the 8.10am Boat of Garten to Craigellachie and Elgin at Carron stations, 21 May 1956. (Transport OnLine Collection)

62277 *Gordon Highlander* on a long local goods train at Craigellachie, 21 May 1956. (Transport OnLine Collection)

Gordon Highlander's name painted on the leading splasher, replacing the brass nameplate fixed when new. (Transport OnLine Collection)

D40 62264, one of the first batch of Pickersgill 'V' class built by Neilson & Co. in 1899, later LNER D40, at Elgin shed, 21 May 1956. (Transport OnLine Collection)

Gresley K2 61783 piloted by D40 62277 *Gordon Highlander* depart from Craigellachie with a heavy goods train, 2 April 1956. (Transport OnLine Collection)

Preserved 49 *Gordon Highlander* at Dawsholm, 18 September 1959. (A.C. Gilbert/MLS Collection)

Preserved 49 *Gordon Highlander* in Great North of Scotland livery (which it never carried), c1960. (A.C. Gilbert/MLS Collection)

D49/1 2759 *Cumberland* at Haymarket shed, August 1937. (Colour Rail)

D49/2 235 *The Bedale* at York, 1938. (Colour Rail)

D49/1 62701 *Derbyshire* at York with a Scarborough–Leeds train, c1954. (Colour Rail)

D49/2 Blaydon's 62771 *The Rufford* at Reedsmouth with a local stopping train to Hexham, 1953. (Colour Rail)

Scarbotough's D49/2
62756 *The Brocklesbury* at Kirkham Abbey with a York–Scarborough semi-fast service, April 1954. (Colour Rail)

Stirling's D49/1
62725 *Inverness-shire* departs from Dalmeny with a Stirling–Edinburgh via Alloa stopping train, June 1957. (Colour Rail)

Colour Section • 207

Selby's D49/1
62731 *Selkirkshire* passes Ulleskelf with a York–Leeds train, May 1958.
(Colour Rail)

D49/1 62712 *Morayshire* before preservation, at Cockburnspath, September 1957.
(Colour Rail)

Preserved D49/1 246 *Morayshire* in LNER lined green livery. (A.C. Gilbert/MLS Collection)

Hornby model R378 62730 *Berkshire* detailed with 'Crownline' brass parts and repainted BR mixed traffic lined black livery, 1981. (David Maidment)

Hornby model R378 converted to a D49/2 with 'Crownline' brass parts and repainted BR mixed traffic lined black as 62751 *The Albrighton*, 1982. (David Maidment)

APPENDIX

The NER class 38
The dimensions are on page 14. All were withdrawn before the Grouping except 281.

Statistics

No.	Built	Worsdell Boiler	Last depot	Withdrawn
38	10/1884	5/1897	Gateshead	2/1915
112	10/1984	9/1898	Middlesbrough	3/1915
126	12/1884	4/1896	York	11/1921
158	12/1884	10/1899	York	10/1921
180	5/1884	8/1896	Malton	10/1919
186	4/1884	12/1895	Selby	3/1915
231	11/1884	8/1898	Selby	1/1915
234	12/1884	1/1897	Leeds	7/1915
281	12/1884	2/1897	York	2/1923
385	8/1884	11/1897	Selby	4/1915
426	6/1884	6/1896	Hull	2/1915
500	9/1884	12/1895	Darlington	2/1917
576	7/1884	6/1896	Gateshead	3/1918
664	6/1884	2/1897	Malton	6/1920
1318	2/1884	8/1895	Darlington	12/1920
1331	9/1884	4/1897	Gateshead	4/1920
1492	10/1884	4/1897	Hull	3/1915
1493	10/1884	4/1896	Hull	4/1916
1494	11/1884	1/1900	Gateshead	5/1915
1495	12/1884	8/1896	Blaydon	1/1915
1496	12/1884	12/1897	Hull	3/1921
1497	1/1885	9.1898	Hull	4/1917
1498	2/1885	5/1899	Hull	1/1915

No.	Built	Worsdell Boiler	Last depot	Withdrawn
1499	2/1885	8/1900	Hull	7/1919
1500	3/1885	11/1896	Middlesbrough	12/1921
1501	3/1885	12/1897	Middlesbrough	11/1920
1502	4/1885	8/1898	Selby	5/1917
1503	4/1895	5/1897	Scarborough	6/1915

The NER Classes M1 & Q (LNER D17/1 & D17/2)

For dimensions, see page 17. All were built at Gateshed.

Statistics

No.	Built	Piston valves	Superheated	1946 No.	Last depot	Withdrawn
D17/1						
1620	12/1892	5/07	10/17	-	Bridlington	2/1934
1621	3/1893	8/05	4/14	(2108)	Alnmouth	7/1945 Preserved
1622	4/1893	6/08	8/15	-	Bridlington	3/1935
1623	5/1893	10/03	6/15	-	Hull	5/1932
1624	6/1893	2/06	11/29	-	Alnmouth	11/1938
1625	6/1893	8/06	1/16	-	Alnmouth	10/1932
1626	6/1893	4/04	2/15	-	Bridlington	5/1934
1627	6/1893	12/07	6/20	-	Alnmouth	12/1933
1628	6/1893	11/06	6/15	-	Bridlington	4/1927 collision damage
1629	6/1893	12/08	6/16	(2109)	Newport	9/1945
1630	6/1893	7/08	11/16	-	Bridlington	12/1933
1631	6/1893	2/04	1/29	-	Selby	7/1934
1632	9/1893	4/04	9/20	-	Bridlington	11/1937
1633	9/1893	9/08	3/14	-	Bridlington	11/1934
1634	10/1893	7/04	8/29	-	Bridlington	1/1935
1635	10/1893	3/08	8/19	-	Bridlington	10/1931
1636	11/1893	5/07	8/20	-	Alnmouth	10/1938
1637	11/1893	9/04	3/21	-	Selby	1/1935
1638	12/1893	12/05	5/24	-	Bridlington	11/1937
1639	3/1894	3/1894	6/14	-	Bridlington	3/1934
D17/2						
1871	6/1896	2/30	2/30	(2110)	Alnmouth	2/1944
1872	6/1896	2/14	2/14	-	W. Hartlepool	11/1933
1873	6/1896	10/13	10/13	2111 3/46	York	2/1948

No.	Built	Piston valves	Superheated	1946 No.	Last depot	Withdrawn
1874	6/1896	10/24	10/28	-	Bridlington	10/1938
1875	9/1896	9/15	9/15	-	Neville Hill	10/1931
1876	9/1896	5/15	5/15	-	Haymarket	4/1943
1877	10/1896	1/16	1/16	-	Alnmouth	10/1938
1878	11/1896	4/16	10/20	-	Scarborough	2/1934
1879	11/1896	7/23	7/25	-	Starbeck	8/1939
1880	12/1896	12/18	12/18	-	W.Hartlepool	1/1933
1901	5/1897	8/19	8/19	(2112)	Duns	6/1945
1902	6/1897	2/20	2/20	2112 1/46	York	2/1948
1903	6/1897	7/27	7/27	-	Starbeck	12/1936
1904	5/1897	9/15	9/15	-	Bridlington	10/1938
1905	6/1897	2/14	2/14	(2114)	Alnmouth	12/1943
1906	6/1897	8/28	8/28	-	Alnmouth	11/1938
1907	6/1897	3/17	3/17	-	Hull	10/1938
1908	6/1897	4/23	4/23	-	Starbeck	11/1938
1909	6/1897	7/17	10/28	-	Bridlington	11/1937
1910	6/1897	8/22	8/22	-	Bridlington	11/1938
1921	6/1897	8/22	8/22	-	St Margaret's	4/1943
1922	9/1897	9/26	10/28	-	Hull	10/1932
1923	9/1897	3/22	5/24	-	Alnmouth	10/1938
1924	10/1897	10/13	10/13	-	Carlisle	10/1937
1925	10/1897	10/19	10/19	-	Manningham	11/1938
1926	10/1897	1/14	1/14	-	Carlisle	8/1931
1927	11/1897	4/14	4/14	-	Starbeck	10/1932
1928	11/1897	3/22	3/22	-	Neville Hill	5/1932
1929	11/1897	8/24	6/31	-	Starbeck	11/1938
1930	11/1897	2/14	2/14	-	W.Hartlepool	11/1933

The NER Q1 (LNER D18)

For dimensions, see page 27. Both were built at Gateshead.

Statistics

No.	Built	Piston valves	Superheated	Last depot	Withdrawn
1869	5/1896	3/20	3/20	Leeds Neville Hill	10/1930
1870	6/1896	9/11	3/15	Leeds Neville Hill	10/1930

The NER 3CC (LNER D19)

For dimensions, see page 30. It was built at Gateshead.

Statistics

No.	Built	Rebuilt	Last depot	Withdrawn
1619	5/1893	8/98	Bridlington	10/1930

The NER R (LNER D20)

For dimensions, see page 31. All were built at Gateshead.

Statistics

No.	Built	Superheated	D20/2	1946 No.	BR No.	Last depot	Withdrawn
2011	8/1899	10/17		2340	62340	Selby	2/1951
2012	9/1899	5/14		2341	62341	Selby	3/1951
2013	9/1899	10/12		2342	62342	Selby	3/1951
2014	10/1899	10/16		2343	62343	Selby	10/1956
2015	11/1899	1/15		2344	62344	Alnmouth	3/1951
2016	11/1899	1/16		2345	62345	Selby	10/1956
2017	11/1899	12/14		(2346)	-	Scarborough	12/1944
2018	12/1899	9/13		2347	62347	Northallerton	11/1954
2019	12/1899	10/15		2348	62348	Selby	2/1951
2020	12/1899	5/14	10/36	2349	62349	Selby	2/1956
2021	8/1900	3/15		2350	-	Selby	12/1947
2022	8/1900	2/13		2351	62351	Alnmouth	11/1954
2023	9/1900	9/21		2352	62352	Gateshead	6/1954
2024	9/1900	1/15		2353	62353	Bridlington	4/1951
2025	9/1900	3/25		2354	62354	Alnmouth	4/1951
2026	10/1900	11/12		2355	62355	Alnmouth	11/1955
2027	10/1900	2/14		(2356)	-	Starbeck	11/1946
2028	11/1900	4/18		2357	62357	Alnmouth	1/1951
2029	12/1900	5/14		2358	62358	Alnmouth	10/1954
2030	12/1900	2/18		2359	62359	Neville Hill	10/1955
2101	12/1900	4/19	12/42	2360	62360	Alnmouth	10/1956
2102	12/1900	6/16		2361	(62361)	Selby	2/1951
2103	12/1900	7/18		2362	62362	Alnmouth	4/1951
2104	2/1901	8/14		2363	62363	Selby	3/1951
2105	3/1901	7/15		2364	-	Starbeck	10/1947
2106	3/1901	2/15		2365	62365	Alnmouth	4/1951

No.	Built	Superheated	D20/2	1946 No.	BR No.	Last depot	Withdrawn
2107	3/1901	4/15		2366	62366	Selby	3/1951
2108	4/1901	11/15		2367	(62367)	Hull	1/1948
2109	4/1901	11/12		(2368)	-	Hull	6/1944
2110	5/1901	9/17		2369	62369	Starbeck	3/1951
476	9/1906	6/15		2370	(62370)	Starbeck	4/1951
592	9/1906	2/15	10/42	2371	62371	Alnmouth	10/1954
707	10/1906	10/17		2372	62372	Bridlington	11/1956
708	10/1906	1/21		2373	62373	Northallerton	2/1953
711	10/1906	5/15		2374	62374	Selby	10/1954
712	11/1906	7/20	10/48	2375	62375	Alnmouth	5/1957
713	11/1906	8/15		2376	(62376)	Selby	2/1951
723	11/1906	2/20		2377	(62377)	Alnmouth	5/1949
724	12/1906	2/15		2378	62378	Selby	11/1956
725	12/1906	8/17		2379	62379	West Hartlepool	4/1951
1026	2/1907	3/14		2380	62380	Gateshead	9/1954
1042	3/1907	6/19		2381	62381	Alnmouth	11/1957
1051	6/1907	2/14		2382	(62382)	Selby	2/1951
1078	6/1907	6/14		2383	62383	Alnmouth	5/1957
1147	3/1907	1/17		-	-	West Hartlepool	1/1943
1184	6/1907	6/14		2384	62384	Selby	8/1955
1206	4/1907	6/17		(2385)	-		1/1945
1207	8/1907	4/13		2386	62386	Selby	10/1956
1209	4/1907	10/17		2387	62387	Alnmouth	9/1957
1210	8/1907	3/14		2388	62388	Northallerton	4/1954
1217	5/1907	12/14		2389	62389	Neville Hill	9/1954
1223	9/1907	12/16		2390	(62390)	Stockton	11/1948
1232	5/1907	3/18		2391	62391	Northallerton	6/1951
1234	5/1907	4/29		-	-	Bridlington	5/1943
1235	9/1907	11/13		2392	62392	Selby	5.1954
1236	6.1907	12/17		2393	-	Starbeck	12/1947
1258	9/1907	11/17		2394	-	Starbeck	5/1946
1260	6/1907	11/14		2395	62395	Alnmouth	11/1957
1665	9/1907	8/20		2396	62396	Alnmouth	11/1957
1672	9/1907	8/13		2397	62397	Bridlington	2/1957

The NER 'R1' (LNER D21)

For dimensions, see page 55. All were built at Darlington. They were allocated 2217–2224 in the LNER 1946 renumbering scheme but none survived long enough to carry the new number.

Statistics

No.	Built	Superheated	Last Depot	Withdrawn
1237	11/1908	7/1915	Neville Hill	3/1945
1238	12/1908	12/1912	Neville Hill	3/1945
1239	12/1908	6/1914	Neville Hill	12/1942
1240	3/1909	4/1915	Neville Hill	5/1944
1241	6/1909	5/1915	Starbeck	3/1943
1242	6/1909	9/1912	Neville Hill	1/1946
1243	6/1909	1/1913	Neville Hill	6/1945
1244	7/1909	8/1912	Neville Hill	10/1944
1245	8/1909	7/1915	Neville Hill	2/1946
1246	8/1909	3/1913	Starbeck	7/1943

The NER 'D', 'F' & 'F1' (LNER D22)

The 'D' 2-4-0

Built as a compound 2-4-0 at Gateshead and rebuilt as a simple expansion 4-4-0 class 'F'. For dimensions see pages 55 & 56.

Statistics

No.	Built	Rebuilt	Superheated	Last depot	Withdrawn
1324	11/1886	10/1896	12/1917	Carlisle	2/1930
340	12/1888	10/1896	8/1915	Hull	12/1933

The 'F' 4-4-0 Compound

Built as a 4-4-0 2 cylinder compound and rebuilt as simple expansion engine with Stephenson valve gear and piston valves as class 'F'. For dimensions see page 57.

Statistics

No.	Built	Rebuilt	Superheated	Last depot	Withdrawn
18	6/1887	3/1905	10/1916	Gateshead	1/1929
42	6/1887	10/1904	2/1915	Scarborough	5/1930
115	6/1887	9/1904	3/1914	Manningham	12/1933
117	11/1887	11/1903	9/1913	Hull	3/1929
355	11/1887	10/1901	9/1915	Carlisle	1/1930
356*	12/1887	12/1903	3/1920	Hull	3/1932 *No.1 1887-1914

No.	Built	Rebuilt	Superheated	Last depot	Withdrawn
514	11/1887	5/1900	3/1914	Hull	2/1929
663	12/1887	1/1900	10/1919	Hull	10/1932
684	12/1887	2/1903	9/1913	Selby	6/1930
779	11/1887	10/1902	8/1914	Scarborough	2/1930
1532	12/1890	9/1905	6/1920	Hull	4/1930
1533	12/1890	7/1905	4/1918	Starbeck	9/1931
1534	12/1890	12/1904	4/1918	Selby	10/1929
1535	12/1890	2/1905	9/1914	Selby	7/1933
1536	12/1890	11/1904	1/1914	Hull	9/1931
1537	12/1890	1/1905	4/1916	Selby	11/1935
1538	12/1890	5/1901	6/1918	Scarborough	6/1933
1539	12/1890	3/1904	12/1913	Hull	2/1930
1540	12/1890	7/1905	7/1918	Hull	10/1929
1541	12/1890	5/1905	2/1920	Starbeck	9/1934
1542	12/1890	6/1905	4/1920	Stockton	7/1933
1543	12/1890	10/1905	1/1917	Hull	8/1932
1544	12/1890	3/1905	2/1916	Selby	9/1929
1545	4/1891	11/1905	3/1916	Starbeck	12/1931
1546	4/1891	4/1901	12/1916	Hull	1/1935

The 'F1' 4-4-0 simple expansion

Built as a simple expansion 4-4-0 as class 'F1' and rebuilt with Stephenson valve gear and piston valves as class 'F'. For dimensions, see page 57.

Statistics

No.	Built	Rebuilt	Superheated	Last depot	Withdrawn	
85	12/1887	2/1903	4/1916	Hull	10/1932	
96	12/1887	9/1903	8/1914	Hull	4/1927	Collision damage
154	12/1887	11/1904	3/1917	Carlisle	2/1930	
194	12/1887	7/1908	12/1915	Hull	4/1930	
230	11/1887	2/1905	2/1918	Hull	2/1930	
673	11/1887	2/1911	2/1916	Selby	10/1933	
777	11/1887	4/1905	11/1914	Waskerley	5/1935	
803	11/1887	6/1911	1/1918	Hull	5/1929	
808	12/1887	5/1905	3/1916	Stockton	2/1930	
1137	12/1887	12/1903	7/1919	Hull	11/1929	

The NER 'G' (LNER D23)

All were built at Darlington, for dimensions, see pages 60 & 61.

Statistics

No.	Built	Rebuilt from 2-4-0	Superheated	Last depot	Withdrawn
23	12/1887	4/1901	2/1913	Waskerley	8/1929
214	10/1888	5/1903	9/1914	Malton	5/1930
217	10/1888	9/1901	4/1914	Middleton-in-Teesdale	8/1931
222	6/1888	8/1901	9/1914	Barnard Castle	12/1930
223	6/1888	6/1901	9/1914	Barnard Castle	4/1933
258	4/1888	3/1902	5/1914	Starbeck	11/1931
274	12/1887	7/1904	3/1916	Darlington	12/1930
328	4/1888	8/1904	6/1914	Manningham	10/1931
337	6/1888	8/1904	2/1914	Barnard Castle	8/1933
372	10/1888	6/1903	4/1913	Darlington	12/1930
472	11/1888	6/1903	9/1913	Darlington	11/1930
521	6/1888	2/1901	11/1914	Darlington	12/1930
557	11/1887	12/1900	1/1913	Darlington	1/1931
675	12/1887	5/1904	3/1914	Barnard Castle	3/1930
676	12/1887	5/1904	4/1914	Barnard Castle	1/1930
677	12/1887	5/1902	11/1913	Kirkby Stephen	12/1933
678	11/1887	7/1902	1/1914	Kirkby Stephen	12/1933
679	12/1887	5/1903	7/1913	Darlington	1/1931
1107	11/1888	12/1901	2/1913	Kirkby Stephen	6/1930
1120	6/1888	12/1902	10/1915	Manningham	5/1935

The Hull & Barnsley Rly 'J' (LNER D24)

For dimensions, see page 62. All were based at Hull throughout their lives.

Statistics

No.	Built	LNER No.	1924 LNER No.	Reboilered	Withdrawn
33	12/1910	3033	2425	9/1929	8/1933
35	12/1910	3035	2426	7/1930	12/1933
38	12/1910	3038	2427	7/1930	1/1934
41	12/1910	3041	2428	4/1930	12/1933
42	12/1910	3042	2429	11/1929	9/1934

The North British 'N' (LNER D25)
For dimensions, see pages 65 & 67. All twelve engines were built at Cowlairs.

Statistics

No.	Built	Reboilered	LNER No.	Last depot	Withdrawn
592	4/1886	6/1911	9592	Eastfield	9/1932
593	9/1886	3/1911	(9593)	Loch Leven	10/1926
594	10/1886	7/1911	(9594)	Eastfield	11/1926
595	4/1887	4/1911	9595	Haymarket	3/1932
596	4/1887	5/1911	9596	Dunfermline	7/1933
597	3/1887	5/1911	(9597)	Haymarket	10/1926
598	2/1888	4/1911	9598	Haymarket	10/1930
599	2/1888	6/1911	9599	Hawick	2/1930
600	2/1888	5/1911	9600	Eastfield	9/1928
601	3/1888	4/1911	9601	Dunbar	4/1926
602	3/1888	4/1911	(9602)	Haymarket	4/1926
603	3/1888	6/1911	9603	Hawick	3/1928

The North British 'K' '317 class (LNER D26)
For dimensions, see page 69. All twelve engines were built at Cowlairs.

Statistics

No.	Built	LNER No.	Last depot	Withdrawn
317	5/1903	(9317)	-	8/1924
318	6/1903	9318	Bathgate	7/1925
319	6/1903	-	-	9/1922
320	6/1903	320B (9320)	Hawick	1/1925
321	7/1903	-	-	11/1922
322	7/1903	9322	Bathgate	2/1925
323	9/1903	(9323)	-	2/1923
324	9/1903	9324	Bathgate	12/1924
325	10/1903	9325	Bathgate	7/1926
326	10/1903	9326	Bathgate	8/1925
327	10/1903	9327	Bathgate	11/1925
328	11/1903	-	-	7/1922

The North British 'M' 'Abbotsford class' (LNER D27 & D28)

For dimensions, see page 71.

Statistics

No.	Built	Name	Builder	Rebuilt NB	1919 No.	LNER No.	Last depot	Withdrawn
476	5/1877	Carlisle	Neilson	6/1902	1321	D17/1 (10321)	Eastfield	11/1924
477	6/1877	Edinburgh	Neilson	7/1904	1322	D17/2 (10322)	Eastfield	12/1924
478	6/1877	Melrose	Neilson	7/1902	1323	D17/1 (10324)	Haymarket	9/1924
479	7/1877	Abbotsford	Neilson	7/1902	1324	D17/1 (10324)	Haymarket	12/1923
486	10/1878	Aberdeen*	Neilson	6/1904	1360	-	-	2/1922
487	10/1878	Montrose*	Neilson	7/1904	1361	D17/2 10361	Haymarket**	9/1926
488	11/1878	Galashiels	Neilson	7/1902	1362	-	-	9/1921
489	11/1878	Hawick	Neilson	7/1902	1363	-	-	6/1922
490	1/1879	St Boswells	Cowlairs	6/1902	1371	-	-	11/1921
491	1/1879	Dalhousie	Cowlairs	7/1904	1387	D17/2 10387	Haymarket	9/1926
492	2/1879	Newcastleton	Cowlairs	7/1904	1388	D17/2 (10388)	Eastfield	12/1924
493	2/1879	Netherby	Cowlairs	6/1904	1389	-	-	6/1921

* 486 renamed *Eskbank* and 487 renamed *Waverley* in 1/1880
** 10361 was seen working around Glasgow in its last few weeks

The North British 'J' 'Scott' class (LNER D29)

For dimensions, see page 74. All twelve engines were built at Cowlairs.

Statistics

No.	Built	Name	LNER No.	Superheated	1946 No.	BR No.	Last depot	Withdrawn
895	7/1909	Rob Roy	9895	4/1925	2400	(62400)		4/1948
896	7/1909	Dandie Dinmont	9896	3/1932	2401	(62401)		11/1949
897	8/1909	Redgauntlet	9897	10/1932	2402	(62402)		6/1949
898	8/1909	Sir Walter Scott	9898	11/1925	2403	(62403)		3/1948
899	8/1909	Jeanie Deans	9899	6/1931	2404	(62404)		8/1949
900	9/1909	The Fair Maid	9900	6/1934	2405	62405	Haymarket	2/1951
243	9/1911	Meg Merrilees	9243	1/1933	2406	(62406)		10/1949
244	10/1911	Madge Wildfire	9244	6/1926	2407	-		12/1947
245	10/1911	Bailie Nicol Jarvie	9245	5/1932	2408	-		11/1947
338	10/1911	Helen Macgregor	9338	8/1935	2409	(62409)		10/1948
339	10/1911	Ivanhoe	9339	6/1931	2410	62410	Thornton Jn	1/1952
340	11/1911	Lady of Avenel	9340	11/1925	2411	62411	Thornton Jn	11/1952
359	12/1911	Dirk Hatteraick	9359	10/1932	2412	62412		9/1950

No.	Built	Name	LNER No.	Superheated	1946 No.	BR No.	Last depot	Withdrawn
360	12/1911	Guy Mannering	9360	10/1930	2413	62413		8/1950
361	12/1911	Vich Ian Vohr	9361	8/1936	(2414)	-		2/1946
362	12/1911	Ravenswood	9362	7/1933	2415	-		7/1947

The North British 'J' 'Superheated Scott' class (LNER D30)

For dimensions, see pages 80 & 81.

Statistics

No.	Built	Name	LNER No.	1946 No.	BR No.	Last depot	Withdrawn
400	9/1912	The Dougal Cratur	9400	(2416)	-	St Margaret's	6/1945
363	10/1912	Hal o' the Wynd	9363	2417	62417	Hawick	1/1951
409	4/1914	The Pirate	9409	2418	62418	Thornton Jn	8/1959
410	4/1914	Meg Dods	9410	2419	62419	Thornton Jn	9/1957
411	4/1914	Dominie Sampson	9411	2420	62420	Hawick	5/1957
412	4/1914	Laird o' Monkbarns	9412	2421	62421	St Margaret's	6/1960
413	5/1914	Caleb Balderstone	9413	2422	62422	Hawick	12/1958
414	6/1914	Dugald Dalgetty	9414	2423	62423	Hawick	12/1957
415	6/1914	Claverhouse	9415	2424	62424	St Margaret's	8/1957
416	6/1914	Ellengowan	9416	2425	62425	Hawick	7/1958
417	7/1914	Cuddie Headrigg	9417	2426	62426	Stirling	6/1960
418	7/1914	Dumbledykes	9418	2427	62427	Dunfermline	4/1959
419	9/1914	The Talisman*	9419	2428	62428	Hawick	12/1958
420	10/1914	The Abbot	9420	2429	62429	Thornton Jn	8/1957
421	10/1914	Jingling Geordie	9421	2430	62430	Thornton Jn	1/1957
422	10/1914	Kenilworth	9422	2431	62431	Thornton Jn	10/1958
423	10/1914	Quentin Durward	9423	2432	62432	Hawick	12/1958
424	6/1915	Lady Rowena	9424	2433	-	Haymarket	11/1947
425	7/1915	Kettledrummle	9425	2434	62434	Dundee	4/1958
426	7/1915	Norna	9426	2435	62435	Hawick	12/1957
427	8/1915	Lord Glenvarloch	9427	2436	62436	Dunfermline	6/1959
428	8/1915	Adam Woodcock	9428	2437	62437	Haymarket	6/1958
497	12/1920	Peter Poundtext	9497	2438	62438	Dundee	10/1957
498	11/1920	Father Ambrose	9498	2439	62439	Bathgate	10/1959
499	11/1920	Wandering Willie	9499	2440	62440	Hawick	7/1958
500	11/1920	Black Duncan	9500	2441	62441	Dunfermline	8/1958
501	12/1920	Simon Glover	9501	2442	62442	Thornton Jn	6/1958

* Just Talisman, 12/1931 - 2/1938 & 5/1940 - 3/1947

The North British Holmes 'M' class, LNER D31

All were constructed at Cowlairs. For dimensions, see pages 93 & 94.

Statistics

No.	Built	Rebuilt	LNER No.	1946 No.	BR No.		Withdrawn
574	6/1884	7/1911	9574	D30/1	-	-	6/1934
575	6/1884	7/1911	9575	D30/1	-	-	2/1937
576	7/1884	7/1911	9576	D30/1	-	-	6/1933
577	7/1884	7/1911	9577	D30/1	-	-	5/1934
578	7/1884	7/1911	9578	D30/1	-	-	7/1933
579	7/1884	7/1911	9579	D30/1	-	-	8/1933
633	5/1890	9/1918	9633	D30/1	-	-	1/1939
634	5/1890	1/1919	9634	D30/1	-	-	11/1935
635	5/1890	10/1918	9635	D30/1	2059	62059, 62281	12/1952
636	6/1890	1/1918	9636	D30/1	-	-	11/1937
637	6/1890	10/1918	9637	D30/1	-	-	4/1936
638	6/1890	9/1918	9638	D30/1	-	-	5/1937
639	8/1890	12/1918	9639	D30/1	-	-	10/1935
640	8/1890	12/1918	9640	D30/1	-	-	10/1936
641	8/1890	10/1918	9641	D30/1	-	-	5/1937
642	9/1890	12/1918	9642	D30/1	2060	62060, 62282	2/1950
36	9/1890	9/1918	9036	D30/1	-	-	6/1935
37	9/1890	12/1918	9037	D30/1	-	-	5/1935
262	11/1894	10/1920	9262	D30/1	-	-	8/1937
293	12/1894	10/1920	9293	D30/1	-	-	1/1938
312	1/1895	1/1921	9312	D30/1	-	-	12/1938
404	1/1895	11/1922	9404	D30/2	2062	(62062)	3/1948
211	1/1895	9/1922	9211	D30/2	-	-	9/1937
212	1/1895	3/1921	9212	D30/1	-	-	4/1937
213	6/1895	10/1920	9213	D30/1	-	-	3/1939
214	6/1895	10/1922	9214	D30/2	-	-	5/1939
215	6/1895	10/1920	9215	D30/1	(2061)	-	9/1946
216	6/1895	11/1920	9216	D30/1	-	-	11/1937
217	7/1895	11/1920	9217	D30/1	-	-	10/1939
218	7/1895	2/1921	9218	D30/1	-	-	7/1937
729	3/1898	12/1921	9729	D30/2	(2063)	-	3/1946

No.	Built	Rebuilt	LNER No.	1946 No.	BR No.		Withdrawn
730	4/1898	12/1920	9730	D30/1	-	-	12/1937
731	4/1898	1/1922	9731	D30/2	2064	(62064)	8/1948
732	5/1898	4/1922	9732	D30/2	2065	62065 (62284)	4/1949
733	5/1898	10/1922	9733	D30/2	2066	(62066)	5/1948
734	5/1898	5/1922	9734	D30/2	2067	-	10/1946
735	7/1898	4/1922	9735	D30/2	-	-	3/1939
736	7/1898	3/1922	9736	D30/2	-	-	7/1931
737	7/1898	12/1920	9737	D30/1	-	-	5/1939
738	7/1898	2/1922	9738	D30/2	-	-	6/1939
739	7/1898	9/1922	9739	D30/2	2068	-	7/1947
740	7/1898	6/1922	9740	D30/2	2069	-	11/1947
765	9/1899	2/1921	9765	D30/1	(2070)	-	11/1946
766	9/1899	6/1922	9766	D30/2	-	-	6/1937
767	9/1899	9/1922	9767	D30/2	2071	-	10/1946
768	11/1899	12/1921	9768	D30/2	2072	62072, 62283	2/1951
769	11/1899	12/1921	9769	D30/2	2073	-	12/1947
770	11/1899	4/1922	9770	D30/2	2074	-	3/1946

The North British 'K' ('Intermediate') Class, LNER D32

All were built at Cowlairs. For dimensions see page 100.

Statistics

No.	Built	Superheated	LNER No.	1946 No.	BR No.	Withdrawn
882	10/1906	8/1924	9882	2443	(62443)	3/1948
883	11/1906	2/1924	9883	2444	(62444)	9/1948
884	11/1906	12/1925	9884	2445	(62445)	12/1949
885	12/1906	11/1923	9885	2446	(62446)	9/1948
886	12/1906	11/1924	9886	2447	-	12/1947
887	12/1906	9/1924	9887	2448	(62448)	9/1948
888	12/1906	8/1925	9888	2449	(62449)	11/1948
889	12/1906	9/1925	9889	2450	(62450)	2/1948
890	1/1907	5/1925	9890	2451	62451	3/1951
891	1/1907	2/1926	9891	2452	-	11/1947
892	1/1907	11/1924	9892	2453	(62453)	5/1948
893	1/1907	1/1925	9893	2454	(62454)	9/1948

The North British 'K' ('Intermediate') Class, LNER D33

All were built at Cowlairs. For dimensions see page 104.

Statistics

No.	Built	Superheated	LNER No.	1946 No.	BR No.	Withdrawn
864	10/1909	12/1928	9864	2455	62455	12/1949
865	10/1909	6/1933	9865	2456	-	12/1947
866	11/1909	4/1936	9866	2457	62457	6/1952
867	11/1909	11/1927	9867	2458	(62458)	9/1949
894	11/1909	11/1925	9894	2459	62459	9/1951
331	10/1909	2/1935	9331	2460	62460	8/1951
332	12/1909	3/1926	9332	2461	62461	6/1951
333	12/1909	3/1934	9333	2462	62462	11/1952
382	12/1909	4/1934	9382	2463	(62463)	3/1948
383	1/1910	7/1933	9383	2464	62464	9/1953
384	2/1910	7/1930	9384	2465	-	9/1947
385	2/1910	11/1935	9385	2466	62466	10/1951

The North British 'K' ('Superheated Intermediate') Class, LNER D34

All were built at Cowlairs. For dimensions see pages 106 & 107.

Statistics

No.	Built	Name	LNER No.	1946 No.	BR No.	Last depot	Withdrawn
149	9/1913	*Glenfinnan*	9149	2467	62467	Thornton Jn	8/1960
221	9/1913	*Glen Orchy*	9221	2468	62468	Thornton Jn	9/1958
256	9/1913	*Glen Douglas*	9256	2469	62469	Dawsholm	12/1962 Preserved
258	9/1913	*Glen Roy*	9258	2470	62470	Perth	5/1959
266	10/1913	*Glen Falloch*	9266	2471	62471	St Margaret's	3/1960
307	12/1913	*Glen Nevis*	9307	2472	62472	Eastfield	10/1959
405	12/1913	*Glen Spean*	9405	2473	(62473)		5/1949
406	12/1913	*Glen Croe*	9406	2474	62474	Eastfield	6/1961
407	12/1913	*Glem Beasdale*	9407	2475	62475	Thornton Jn	6/1959
408	12/1913	*Glen Sloy*	9408	2476	(62476)		2/1950
100	5/1917	*Glen Dochart*	9100	2477	62477	Eastfield	10/1959
291	5/1917	*Glen Quoich*	9291	2478	62478	Thonton Jn	12/1959
298	5/1917	*Glen Sheil*	9298	2479	62479	Kittybrewster	6/1961
153	6/1917	*Glen Fruin*	9153	2480	62480	Kittybrewster	10/1959
241	7/1917	*Glen Ogle*	9241	2481	(62481)		9/1949

No.	Built	Name	LNER No.	1946 No.	BR No.	Last depot	Withdrawn
242	3/1919	Glen Mamie	9242	2482	62482	Kittybrewster	3/1960
270	3/1919	Glen Garry	9270	2483	62483	Hawick	4/1959
278	4/1919	Glen Lyon	9278	2484	62484	Perth	11/1961
281	4/1919	Glen Murran	9281	2485	62485	Dunfermline	3/1960
287	4/1919	Glen Gyle	9287	(2486)	-		2/1946
503	5/1920	Glen Arklet	9503	2487	62487	St Margaret's	9/1959
504	4/1920	Glen Aladale	9504	2488	62488	Hawick	10/1960
490	5/1920	Glen Dessary	9490	2489	62489	Kittybrewster	12/1959
502	5/1920	Glen Fintaig	9502	2490	62490	St Margaret's	2/1959
505	5/1920	Glen Cona	9505	2491	-		12/1947
34	6/1920	Glen Garvin	9034	2492	62492	Thornton Jn	6/1959
35	6/1920	Glen Gloy	9035	2493	62493	Kittybrewster	6/1960
492	7/1920	Glen Gour	9492	2494	62494	Hawick	4/1959
493	7/1920	Glen Luss	9493	2495	62495	Bathgate	4/1961
494	8/1920	Glen Loy	9494	2496	62496	Eastfield	11/1961
495	8/1920	Glen Mallie	9495	2497	62497	Kittybrewster	2/1960
496	9/1920	Glen Moidart	9496	2498	62498	Kittybrewster	3/1960

The North British 'N' Class, LNER D35

All were built at Cowlairs. For dimensions see page 116.

Statistics

No.	Built	Rebuilt	New NB No.	LNER No.	LNER 1924 No.	Withdrawn
55	5/1894	-	1434	(10434)	-	3/1923
227	1/1896	-	1435	-	-	1920
231	1/1896	-	1436	-	-	12/1922
232	1/1896	-	1437	-	-	10/1921
341	1/1896	-	1438	-	-	9/1921
342	1/1896	-	1439	(10439)	(9997)	11/1924
343	2/1896	-	1440	-	-	9/1921
344	3/1896	-	1441	-	-	7/1921
345	4/1896	-	1442	(10442)	-	2/1923
346	4/1896	-	1443	-	-	10/1921
394	5/1894	-	1444	-	-	6/1922
395	5/1894	-	1445	-	-	5/1922
693	1/1894	-	1446	-	-	9/1921

No.	Built	Rebuilt	New NB No.	LNER No.	LNER 1924 No.	Withdrawn
694	1/1894	-	1447	-	-	9/1921
695	1/1894	2/1919	-	-	-	To class D36
696	1/1894	-	1448	(10448)	(9998)	10/1924
697	1/1894	-	1449	(10449)	-	5/1923
698	1/1894	-	1433	-	-	1920
699	4/1894	-	1450	-	-	11/1922
700	4/1894	-	1451	-	-	4/1921
701	4/1894	-	1452	(10452)	-	5/1923
702	2/1896	-	1398	-	-	1919
703	2/1896	-	1399	-	-	1919
704	3/1896	-	1453	(10453)	-	1/1924

The North British 'L' Class, LNER D36
For dimensions see page 117.

Statistics

No.	Built	Rebuilt	LNER No.	Reboilered (Saturated).	Withdrawn
695	1/1894*	2/1919 & 9/1925	9695	5/1936	5/1943

* as NBR Class 'N'

The Great North of Scotland 'Q' Class, LNER D38
For dimensions see page 119.

Statistics

No.	Built	Builder	LNER No.	Superheated	Withdrawn
75	8/1890	R.Stephenson & Co.	6875	7/1917	1/1938
76	8/1890	R.Stephenson & Co.	6876	-	2/1931
77	9/1890	R.Stephenson & Co.	6877	10/1913	9/1937

The Great North of Scotland 'C' Class, LNER D39
For dimensions see page 121.

Statistics

No.	Built	Builder	Rebuilt	LNER No.	Withdrawn
1	12/1878	Neilson & Co.	2/1897	-	8/1925
2	1/1879	Neilson & Co.	10/1904	6802	6/1926
3	1/1879	Neilson & Co.	2/1898	6803	2/1927

The Great North of Scotland 'V & F' Classes, LNER D40

For dimensions see page 122.

Statistics

No.	Built	Builder	Superheated	LNER No.	1946 No.	BR No.	Withdrawn
25	10/1899	Neilson & Co.	-	6825	2260	62260	8/1953
26	10/1899	Neilson & Co.	-	6826	2261	62261	2/1953
113	10/1899	Neilson & Co.	-	6913	2262	62262	10/1955
114	10/1899	Neilson & Co.	-	6914	(2263)	-	3/1946
115	10/1899	Neilson & Co.	-	6915	2264	62264	3/1957
27	4/1909	Inverurie	-	6827	2265	62265	12/1956
28	3/1913	Inverurie	-	6828	2266	-	1/1947
29	7/1909	Inverurie	-	6829	2267	62267	8/1956
31	6/1910	Inverurie	-	6831	2268	62268	7/1956
33	9/1913	Inverurie	-	6833	2269	62269	9/1955
34	3/1915	Inverurie	-	6834	2270	62270	9/1953
35	9/1914	Inverurie	-	6835	2271	62271	11/1956
36	8/1910	Inverurie	-	6836	2272	62272	3/1955
45 *George Davidson*	6/1921	Inverurie	6/1921	6845	2273	62273	1/1955
46 *Benachie*	9/1921	Inverurie	9/1921	6846	2274	62274	9/1955
47 *Sir David Stewart*	9/1920	NB Loco Co.	9/1920	6847	2275	62275	12/1955
48 *Andrew Bain*	10/1920	NB Loco Co.	10/1920	6848	2276	62276	8/1955
49 *Gordon Highlander*	10/1920	NB Loco Co.	10/1920	6849	2277	62277	6/1958 Preserved
50 *Hatton Castle*	10/1920	NB Loco Co.	10/1920	6850	2278	62278	7/1955
52 *Glen Grant*	10/1920	NB Loco Co.	10/1920	6852	2279	62279	5/1955
54 *Southesk*	10/1920	NB Loco Co.	10/1920	6854	2280	-	1/1947

The Great North of Scotland 'S & T' Classes, LNER D41

For dimensions see page 132. All built by Neilson & Co.

Statistics

No.	Built	GNoS Class	LNER No.	1946 No.	BR No.	Withdrawn
78	12/1893	S	6878	2225	62225	2/1953
79	12/1893	S	6879	(2226)	-	7/1946
80	12/1893	S	6880	2227	62227	3/1951
81	12/1893	S	6881	2228	62228	2/1952
82	12/1893	S	6882	2229	62229	12/1951
83	12/1893	S	6883	2230	62230	3/1952
93	12/1895	T	6893	2237	-	12/1946
94	12/1895	T	6894	2238	62238	8/1948
95	12/1895	T	6895	2239	-	12/1947
96	12/1895	T	6896	2240	62240	10/1949
97	12/1895	T	6897	2241	62241	2/1953
98	12/1895	T	6898	2242	62242	2/1953
99	12/1895	T	6899	2243	62243	1/1951
100	2/1896	T	6900	2244	-	7/1947
19	2/1896	T	6819	2231	62231	11/1952
20	2/1896	T	6820	2232	62232	10/1951
21	2/1896	T	6821	(2233)	-	9/1946
22	2/1896	T	6822	2234	62234	11/1949
23	3/1896	T	6823	2235	(62235)	5/1950
24	3/1896	T	6824	2236	-	12/1947
101	9/1897	T	6901	2245	-	12/1947
102	9/1897	T	6902	2246	62246	8/1951
103	9/1897	T	6903	2247	62247	10/1950
104	9/1897	T	6904	2248	62248	10/1952
105	9/1897	T	6905	2249	62249	10/1950
106	9/1897	T	6906	2250	-	12/1947
107	2/1898	T	6907	2251	62251	6/1951
108	2/1898	T	6908	2252	62252	11/1951
109	2/1898	T	6909	2253	-	1/1947
110	2/1898	T	6910	2254	-	1/1947
111	2/1898	T	6911	2255	62255	5/1952
112	2/1898	T	6912	2256	62256	12/1952

The Great North of Scotland 'O' Class, LNER D42
For dimensions see page 141. All built by Kitson & Co.

Statistics

No.	Built	Superheated	LNER No.	1946 No.	Withdrawn
4	5/1888	-	6804	-	4/1935
7	5/1888	-	6807	(2075)	4/1945
9	5/1888	-	6809	-	11/1939
10	4/1888	-	6810	-	11/1939
17	4/1888	1/1920	6817	(2076)	2/1946
18	5/1888	7/1920	6818	-	12/1939
72	6/1888	12/1920	6872	-	12/1938
73	6/1888	1/1920	6873	-	4/1937
74	6/1888	4/1916	6874	-	5/1939

The Great North of Scotland 'P' Class, LNER D43
For dimensions see page 144. All built by Robert Stephenson & Co.

Statistics

No.	Built	Superheated	LNER No.	Withdrawn
12	5/1890	5/1917 – 6/1932	6812	1/1938
13	6/1890	-	6813	6/1937
14	6/1890	10/1917 – 1/1928 & 2/1931 – 11/1936	6814	11/1936

The Great North of Scotland 'A' Class, LNER D44
For dimensions see page 146. All built by Kitson & Co.

Statistics

No.	Built	Rebuilt	LNER No.	Withdrawn
63	8/1884	10/1905	(6863)	7/1924
64	8/1884	10/1905	(6864)	11/1925
65	8/1884	1/1912	6865	8/1926
66	9/1884	4/1912	(6866)	11/1925
67	9/1884	6/1906	6867	10/1932
68	10/1884	7/1906	(6868)	1/1925

The Great North of Scotland 'M' Class, LNER D45
For dimensions see page 147. All built by Neilson & Co.

Statistics

No.	Built	Rebuilt	LNER No.	Withdrawn
40	10/1878	10/1896	6840	6/1932
51	11/1878	1/1899	6851	1/1927
53	11/1878	10/1903	6853	3/1927
57	8/1878	1/1900	(6857)	6/1925
58	8/1878	11/1903	6858	5/1927
59	8/1878	9/1898	6859	7/1926
60	9/1878	4/1897	(6860)	9/1925
61	9/1878	3/1900	6861	4/1926
62	10/1878	5/1904	(6862)	6/1926

The Great North of Scotland 'N' Class, LNER D46
For dimensions see page 148. All built at Kittybrewster.

Statistics

No.	Built	Rebuilt	LNER No.	Withdrawn
5	2/1887	4/1915	6805	4/1936
6	12/1887	8/1917	6806	2/1932

The Great North of Scotland 'K & L' Classes, LNER D47
For dimensions see pages 149 & 150. All built by Neilson & Co.

Statistics

No.	Built	Class	Rebuilt	GNoS	Dupl list	LNER No.	Withdrawn
43	3/1866	K	5/1890	43A	4/1916	-	8/1921
44	3/1866	K	11/1890	44A	4/1916	-	6/1925
45	3/1866	K	2/1891	45A	7/1921	-	7/1925
46	3/1866	K	2/1890	-		-	7/1921
47	4/1866	K	8/1890	47A	10/1920	-	8/1921
48	4/1866	K	11/1889	48A	10/1920	-	6/1925
49	3/1876	L	8/1898	49A	10/1920	-	10/1924
50	3/1876	L	7/1899	50A	10/1920	-	10/1924
54	3/1876	L	8/1897	54A	11/1920	-	10/1924
55	4/1876	L	3/1900	-		(6855)	8/1924
56	4/1876	L	12/1901	-		(6856)	9/1924
57*	4/1876	L	6/1897	52A	10/1920	-	1/1926

* renumbered 52 in 1878.

The Great North of Scotland 'G' Class, LNER D48

For dimensions see page 151. All built by Kitson & Co.

Statistics

No.	Built	Rebuilt	LNER No.	Withdrawn
69	5/1885	12/1905	6869	11/1934
70	5/1885	9/1906	6870	6/1928
71	5/1885	12/1911	6871	1/1928

The LNER D49/1 and D49/2

For dimensions see pages 152 & 157. All built at Darlington.

Statistics

No.	Name	Built	1946 No.	BR No	Last depot	Withdrawn
D49/1						
234	*Yorkshire*	10/1927	2700	62700	Botanical Gardens	10/1958
251	*Derbyshire*	11/1927	2701	62701	Dairycoates	9/1959
253	*Oxfordshire*	11/1927	2702	62702	Neville Hill	11/1958
256	*Hertfordshire*	12/1927	2703	62703	Botanical Gardens	6/1958
264	*Stirlingshire*	12/1927	2704	62704	Thornton Jcn	8/1958
265	*Lanarkshire*	12/1927	2705	62705	Haymarket	11/1959
266	*Forfarshire*	12/1927	2706	62706	Thornton Jcn	2/1958
236	*Lancashire*	1/1928	2707	62707	Dairycoates	10/1959
270	*Argyllshire*	1/1928	2708	62708	Thornton Jcn	5/1959
277	*Berwickshire*	1/1928	2709	62709	Haymarket	1/1960
245	*Lincolnshire*	2/1928	2710	62710	Dairycoates	10/1960
281	*Dumbartonshire*	2/1928	2711	62711	St Margaret's	5/1961
246	*Morayshire*	2/1928	2712	62712	Hawick	7/1961 Preserved
249	*Aberdeenshire*	2/1928	2713	62713	Thornton Jcn	9/1957
250	*Perthshire*	3/1928	2714	62714	Stirling	8/1959
306	*Roxburghshire*	3/1928	2715	62715	St Margaret's	6/1959
307	*Kincardineshire*	3/1928	2716	62716	St Margaret's	4/1961
309	*Banffshire*	3/1928	2717	62717	Dairycoates	1/1961
310	*Kinross-shire*	5/1928	2718	62718	St Margaret's	4/1961
311	*Peebles-shire*	5/1928	2719	62719	Hawick	1/1960

No.	Name	Built	1946 No.	BR No	Last depot	Withdrawn
D49/3 (oscillating cam poppet valves), rebuilt D49/1 1938						
318	*Cambridgeshire*	5/1928	2720	62720	Dairycoates	10/1959
320	*Warwickshire*	5/1928	2721	62721	St Margaret's	8/1958
322	*Huntingdonshire*	7/1928	2722	62722	Dairycoates	10/1959
327	*Nottinghamshire*	7/1928	2723	62723	Dairycoates	1/1961
335	*Bedfordshire*	8/1928	2724	62724	Botanical Gardens	12/1957
329	*Inverness-shire*	8/1928	2725	62725	Stirling	11/1958
D49/2						
352	*Leicestershire/The Meynell (6/32)*	3/1929	2726	62726	Scarborough	12/1957
336	*Buckinghamshire/The Quorn (5/32)*	6/1929	2727	62727	Dairycoates	1/1961
D49/1						
2753	*Cheshire*	2/1929	2728	62728	Thornton Jcn	10/1959
2754	*Rutlandshire*	4/1929	2729	62729	St Margaret's	5/1961
2755	*Berkshire*	3/1929	2730	62730	Selby	12/1958
2756	*Selkirkshire*	3/1929	2731	62731	Selby	4/1959
2757	*Dumfries-shire*	3/1929	2732	62732	Carlisle Canal	11/1958
2758	*Northumberland*	3/1929	2733	62733	St Margaret's	4/1961
2759	*Cumberland*	5/1929	2734	62734	Carlisle Canal	3/1961
2760	*Westmorland*	6/1929	2735	62735	Scarborough	8/1958
D49/2						
201	*The Bramham Moor*	4/1932	2736	62736	Starbeck	6/1958
211	*The York and Ainsty*	5/1932	2737	62737	Botanical Gardens	1/1958
220	*The Zetland*	5/1932	2738	62738	Starbeck	9/1959
232	*The Badsworth*	5/1932	2739	62739	Scarborough	10/1960
235	*The Bedale*	6/1932	2740	62740	Dairycoates	8/1960
247	*The Blankney*	7/1932	2741	62741	Botanical Gardens	10/1958
255	*The Braes of Derwent*	8/1932	2742	62742	Neville Hill	11/1958
269	*The Cleveland*	8/1932	2743	62743	Haymarket	5/1960
273	*The Holderness*	10/1932	2744	62744	Hawick	12/1960
282	*The Hurworth*	10/1932	2745	62745	Scarborough	3/1959
283	*The Middleton*	8/1933	2746	62746	Starbeck	5/1958
288	*The Percy*	8/1933	2747	62747	Carlisle Canal	3/1961

No.	Name	Built	1946 No.	BR No	Last depot	Withdrawn
292	The Southwold	8/1933	2748	62748	Neville Hill	12/1957
297	The Cottesmore	8/1933	2749	62749	Neville Hill	7/1958
298	The Pytchley	9/1933	2750	62750	Botanical Gardens	11/1958
205	The Albrighton	7/1934	2751	62751	Scarborough	3/1959
214	The Atherstone	7/1934	2752	62752	Starbeck	7/1958
217	The Belvoir	7/1934	2753	62753	Starbeck	9/1959
222	The Berkeley	7/1934	2754	62754	Botanical Gardens	11/1958
226	The Bilsdale	7/1934	2755	62755	Selby	11/1958
230	The Brocklesby	8/1934	2756	62756	Scarborough	4/1958
238	The Burton	8/1934	2757	62757	Botanical Gardens	12/1957
258	The Cattisbrook	8/1934	2758	62758	Starbeck	12/1957
274	The Craven	8/1934	2759	62759	Dairycoates	1/1961
279	The Cotswold	9/1934	2760	62760	Dairycoates	10/1959
353	The Derwent	9/1934	2761	62761	Selby	12/1957
357	The Fernie	9/1934	2762	62762	Scarborough	10/1960
359	The Fitzwilliam	9/1934	2763	62763	Dairycoates	1/1961
361	The Garth	10/1934	2764	62764	Scarborough	11/1958
362	The Goathland	10/1934	2765	62765	Dairycoates	1/1961
363	The Grafton	11/1934	2766	62766	Botanic Gardens	9/1958
364	The Grove	11/1934	2767	62767	Botanic Gardens	10/1958
365	The Morpeth	12/1934*	2768	62768	Starbeck	11/1952
	* rebuilt as 2-cyl class 'D', 8/1942					
366	The Oakley	12/1934	2769	62769	Scarborough	9/1958
368	The Puckeridge	12/1934	2770	62770	Scarborough	9/1959
370	The Rufford	1/1935	2771	62771	York	10/1958
374	The Sinnington	1/1935	2772	62772	Selby	9/1958
375	The South Durham	1/1935	2773	62773	Neville Hill	8/1958
376	The Staintondale	2/1935	2774	62774	Neville Hill	11/1958
377	The Tynedale	2/1935	2775	62775	Selby	12/1958

BIBLIOGRAPHY

Becket, W.S., *The Xpress Locomotive Register, Volume 3, Eastern, North Eastern & Scottish (Ex LNER) Regions,* Xpress Publishing

Hoole, K., *The 4-4-0 classes of the North Eastern Railway,* Ian Allen, 1979

Maidment, David J., *LNER 4-6-0 Locomotive Classes,* Pen & Sword, 2021

Nock, O.S., *LNER Steam,* David & Charles, 1969

RCTS, *Locomotives of the LNER Part 3C, Tender Engines – classes D13-D24,* RCTS 1981

RCTS, *Locomotives of the LNER Part 4, Tender Engines – classes D25–E7,* RCTS 1968

Tuplin, W.A., *North Eastern Steam,* George Allen & Unwin Ltd., 1970

Yeadon's *Register of LNER Locomotives Vol. 34 – Class D17–D24,* Book Law Publications, 2004

Yeadon's *Register of LNER Locomotives Vol. 40 – Class D25–D30,* Book Law Publications, 2007

Yeadon's *Register of LNER Locomotives Vol. 42 – Class D31–D36,* Book Law Publications, 2007

INDEX

Engineers
Chalmers, Walter, 11
Cowan, William, 11
Drummond, Dugald, 10-11
Gresley, Nigel, 12-13
Johnson, James, 12
McDonnell, Alexander, 9
Manson, James, 11-12
Pickersgill, William, 12
Reid, William, 11
Stirling. Matthew, 10
Thompson, Edward, 13
Worsdell, Thomas, 9
Worsdell, Wilson, 9-10

Locomotives
Great North of Scotland Railway
'A' class (LNER D44)
    Construction, 146
    Dimensions, 146
    Numbers, 146
    Operation, 146, 147
    Rebuilding, 146
    Renumbering, 146
    Statistics, 227
    Tenders, 146
    Withdrawal, 146
'C' class (LNER D39)
    Construction, 121
    Dimensions, 121
    Numbers, 121
    Operation, 122
    Rebuilding, 121
    Statistics, 224
    Withdrawal, 121, 122
'G' class (LNER D48)
    Allocation, 151
    Construction, 151
    Dimensions, 151
    Numbers, 151
    Operation, 151
    Rebuilding, 151
    Renumbering, 151
    Statistics, 229
    Tenders, 151
    Withdrawal, 151
'K' & 'L' classes (LNER D47)
    Allocation, 150
    Construction, 149, 150
    Dimensions, 149, 150
    Livery, 150
    Numbers, 149, 150
    Operation, 150
    Renumbering, 149, 150
    Statistics, 228
    Withdrawal, 149-151
'M' class (LNER D45)
    Allocation, 147
    Construction, 147
    Dimensions, 147
    Numbers, 147
    Operation, 147
    Renumbering, 147
    Rebuilding, 147
    Statistics, 228
    Withdrawal, 147
'N' class (LNER D46)
    Construction, 147
    Dimensions, 148
    Names, 148
    Numbers 148
    Operation, 148
    Rebuilt, 148
    Renumbering, 148
    Statistics, 228
    Withdrawal, 148
'O' class (LNER D42)
    Allocation, 142
    Construction, 141
    Dimensions, 141
    Numbers, 141
    Operation, 142
    Renumbering, 141
    Statistics, 227
    Superheating, 141
    Tenders, 141
    Withdrawal, 141
'P' class (LNER D43)
    Allocation, 146
    Construction, 144
    Dimensions, 144
    Numbers, 144
    Operation, 146
    Statistics, 227
    Superheating, 144
    Withdrawal, 144
'Q' class (LNER D38)
    Allocation, 121
    Construction, 119, 120
    Costs, 119
    Dimensions, 119
    Livery, 121
    Numbers, 119, 120
    Operation, 121
    Renumbering, 120
    Statistics, 224
    Superheating, 120
    Tenders, 121
    Withdrawal, 121
'S' & 'T' classes (LNER D41)
    Allocation, 138

Construction, 132
Dimensions, 132
Extended smokeboxes, 132
Numbers, 132
Operation, 137, 138
Renumbering, 132, 135
Statistics, 226
Tenders, 135
Withdrawal, 135

'V' & 'F' classes (LNER D40)
Construction, 122, 123
Dimensions, 122, 123
Livery, 125
Names, 124, 125
Numbers, 122
Operation, 126, 128
Performance, 128, 129
Preservation, 125, 189, 190
Renumbering, 123-125
Statistics, 225
Superheating, 123
Withdrawal, 122

LNER D49 Shire & Hunt classes
D49/1 'Shire' class
    Allocation, 162, 164, 166, 171-173, 177, 178
    Construction, 152, 155
    Design, 152
    Dimensions, 152, 155
    Lentz valve gear, 153
    Livery, 158
    Lubrication, 157
    Names, 153, 155
    Numbers, 152, 153
    Operation, 162, 177
    Performance, 164, 167, 168, 172, 177
    Preserved, 191, 192
    Rebuilding, 155
    Renumbering, 158
    Rough riding, 158, 172, 173
    Statistics, 229, 230
    Steaming problems, 157
    Tenders, 157
    Withdrawals, 178
D49/2 'Hunt' class
    Allocation, 164, 166, 171, 173, 178

Construction, 155
Kylchap exhaust, 157, 158
Livery, 158
Lubrication, 157
Names, 155-157
Numbers, 155-157
Operation, 166, 177
Performance, 167, 168, 171, 172
Rebuilding 365 as class D, 158
Renumbering, 158
Statistics, 230, 231
Tenders, 157
Tests, 157, 158
Withdrawals, 178

North British Railway
'574' class (also 'M', LNER D31)
    Allocation, 99, 100
    Construction, 93, 94
    Dimensions, 93, 94
    Livery, 95
    Numbers, 93, 94
    Operations, 97, 99, 100
    Performance, 99
    Rebuilding, 94
    Renumbering, 94, 95
    Statistics, 220, 221
    Withdrawals, 94, 95, 100
'J' class (LNER D29)
    Allocation, 78, 79
    Construction, 74
    Dimensions, 74
    Livery, 75
    Names, 74, 75
    Numbers, 74
    Operation, 76, 78, 80
    Renumbering 75
    Statistics, 218, 219
    Superheating, 74, 75
    Withdrawal, 75, 80
'J' class (LNER D30)
    Allocation, 84, 85, 87, 91
    Construction, 80
    Dimensions, 80, 81
    Livery, 81
    Names, 81
    Numbers, 81
    Operation, 84, 85

Performance, 88
Renumbering, 80, 82
Statistics, 219
Superheating, 81, 82
Withdrawals, 82, 91, 92
'K' class (LNER D26)
    Construction, 69
    Design, 69,
    Dimensions, 69
    Numbers, 69
    Operation, 71
    Statistics, 217
    Withdrawn, 69, 71
'K' class (LNER D32)
    Allocation, 102, 103
    Construction, 100
    Dimensions, 100, 101
    Livery, 101
    Lubrication, 101
    Numbers, 100
    Operation, 102, 103
    Performance, 102, 103
    Renumbering, 101, 102
    Statistics, 221
    Superheating, 101
    Withdrawal, 102
'K' Intermediate class (LNER D33)
    Allocation, 105, 106
    Construction, 104
    Dimensions, 104
    Livery, 104
    Numbers, 104
    Operation, 105, 106
    Renumbering, 104
    Statistics, 222
    Superheating, 104
    Tests, 106
    Withdrawals, 105, 106
'K' Intermediate Superheated class 'Glens', (LNER D34)
    Allocation, 109, 110, 113, 114
    Construction, 106
    Dimensions, 106, 107
    Lubrication, 108
    Livery, 108
    Names, 107, 108
    Numbers, 106
    Operation, 109, 110, 113

Index • 235

Preserved, 187-189
Renumbering, 108
Statistics, 222, 223
Superheaters, 106, 108
Withdrawal, 108, 109, 114
'L' class (LNER D36)
   Construction, 117
   Dimensions, 117
   Mileage, 118
   Operation, 118
   Statistics, 224
   Withdrawal, 118
'M' class (LNER D27/28)
   Allocation, 72
   Constructed, 71
   Dimensions, 71, 72
   Names, 71
   Numbers, 71
   Operation, 72, 74
   Performance, 73
   Rebuilt to D28, 71
   Renumbering, 72, 74
   Statistics, 218
   Withdrawal, 72, 74
'N' class (LNER D25)
   Allocation, 68, 69
   Construction, 65
   Dimensions, 65, 67
   Numbers, 65, 67
   Operation, 68, 69
   Statistics, 217
   Withdrawal, 68
'N' class (LNER D35)
   Allocation, 116, 117
   Construction, 116
   Dimensions, 116
   Numbers, 116
   Operation, 116
   Performance, 116, 117
   Renumbering, 116
   Statistics, 223, 224
   Withdrawals, 116

North Eastern Railway
'38' class (1884)
   Allocation, 17
   Complaints, 15
   Construction, 14

   Cost, 14
   Dimensions, 14
   Numbers, 14
   Operation, 15
   Reboilering, 15
   Statistics, 209, 210
   Tenders, 15
   Tests, 15
   Withdrawal, 15
'3CC' class (LNER D19)
   Construction, 28
   Cost, 28
   Dimensions, 28, 30
   Operation, 30
   Rebuilding, 28
   Statistics, 212
   Tests, 28
   Withdrawal, 30
'D' & 'F' (LNER D22)
   Accidents, 57
   Allocation, 58, 59
   Compound comparison, 56
   Construction, 55, 56
   Design, 55
   Dimensions, 55
   Livery, 57
   Numbers, 56
   Operation, 57
   Performance, 56, 57
   Rebuilding, 58
   Statistics, 214, 215
   Withdrawal, 57, 59
'G' (LNER D23)
   Allocation, 61, 62
   Construction, 59
   Dimensions, 60, 61
   Livery, 61
   Numbers, 59
   Operation, 61
   Performance, 61
   Rebuilding, 61
   Statistics, 216
   Withdrawal, 61, 62
'J' (LNER D24)
   Allocation, 64
   Construction, 62
   Dimensions, 62
   Livery, 63

   Numbers, 62
   Operation, 64
   Reboilering, 63
   Renumbering, 63
   Statistics, 216
   Withdrawal, 64
'M' & 'Q' (LNER D17)
   Accidents, 23
   Allocation, 20, 22, 23, 27
   Construction, 17
   Cost, 17
   Design, 17
   Dimensions, 17, 18
   Livery, 18
   Operation, 20, 22, 25
   Performance, 20, 22, 25
   Preserved, 186, 187
   Renumbering, 20
   Statistics, 210, 211
   Tenders, 17
   Withdrawal, 20, 27
'Q1' (LNER D18)
   Allocation, 28
   Construction, 27
   Dimensions, 27
   Livery, 28
   Operation, 28
   Performance, 28
   Statistics, 211
   Withdrawal, 28
'R' (LNER D20)
   Accidents, 39
   Allocation, 36, 41, 42, 45-47
   Construction, 30
   Dimensions, 30, 31
   Livery, 31, 32
   Mileage, 36
   Numbers, 30, 31, 35
   Operation, 36, 42, 45
   Performance, 36, 39, 40, 47
   Rebuilding, 32, 35, 36
   Statistics, 212-214
   Superheating, 31
   Tenders, 35
   Tests, 40
   Withdrawal, 35, 36, 47'
R1' (LNER D21)
   Accidents, 52, 55

Allocation, 50, 52, 55
Construction, 49
Dimensions, 49, 57
Livery, 49, 57
Mileage, 55
Numbers, 49, 55
Operation, 50
Performance, 51, 52
Statistics, 214
Superheating, 49
Tests, 52
Withdrawal, 49, 55

Logs
Aberdeen – Craigellachie, 128
Aberdeen – Keith, 137
Aberdeen – Dundee – Edinburgh, 86, 87, 90
Darlington – York, 21, 37-39, 52, 165
Edinburgh – Galashiels – Carlisle, 102
Edinburgh – Newcastle, 40, 51, 162-163
Elgin – Huntley, 127
Fort William – Craigendoran - Helensburgh, 111-112
Glasgow – Edinburgh, 73, 79, 80, 85, 88, 167
Harrogate – Darlington, 41
Leeds – Selby - Hull, 25, 45, 46, 54
Leeds – Harrogate, 44
Newcastle – Chevington, 23
Perth – Dunfermline, 89
Stirling – Edinburgh, 92
York – Bridlington, 61
York – Leeds, 24, 43, 53, 168-169
York – Scarborough, 43
York – Newcastle, 51

Photographs (B&W)
Locations
Aberdeen, 99, 189
Aberdour, 70, 71, 76-78, 98, 116
Alnmouth, 172
Ardlui, 110
Ballater, 128, 131
Balloch, 95, 114
Banff, 145

Benton Quarry, 58
Boat of Garten, 138
Buckie, 140
Carlisle, 26, 70, 96, 100, 115
Carlisle Canal, 96
Carlisle London Road, 46
Chathill, 38
Church Fenton, 175
Cornhill, 151
Craigellachie, 131
Cross Gates, 47, 180
Cudworth, 48
Dalgetty, 71
Darlington, 18, 29, 34, 49, 50,60
Darlington North Road, 36
Derby, 15
Doncaster, 153
Dundee Esplanade, 93
Dundee Tay Bridge, 160
Dunfermline, 83, 106
Edinburgh area, 73, 74
Edinburgh Princes St Gardens, 65, 89, 92, 102, 166-168, 182
Edinburgh St Margarets, 76, 108, 109, 185
Edinburgh Waverley, 56, 182, 183
Elgin, 131, 133, 142, 143
Eryholme, 54
Escrick, 176
Fort William, 107, 108
Forth Bridge, 168
Fountainhall Junction, 191
Fraserburgh, 139
Galashiels, 103
Gateshead, 15
Glasgow Cowlairs, 154
Glasgow Eastfield, 109
Glenfarg, 91, 188
Glenfinnan, 110
Grantshouse, 179
Greatham, 171
Hawes, 48
Hawick, 92, 107
Haymarket, 67, 75, 78, 79, 83, 105, 160
Horsforth, 180
Hull Botanic Gardens, 19, 30, 32, 57
Hull Paragon, 178
Huntley, 136

Inverkeithing, 95, 105184, 185
Inverurie, 129, 139
Keith, 120, 125, 134, 136
Kinghorn, 117
Kirkham Abbey, 25, 42
Kittybrewster, 95, 114, 124, 125, 129, 134, 145-148, 189
Leeds, City, 48, 179
Leeds, Holbeck, 159
Leeds Neville Hill, 29, 158
Leeds Starbeck, 34, 158, 161
Loch Leven, 97
Lossiemouth, 140, 143
Macduff, 130
Mallaig, 113
Malton, 175
Maud, 126
Middleton-in-Teesdale, 60, 62
Perth, 104, 109, 113, 174
Scarborough, 57, 156, 161, 173
Selby, 50
Selkirk, 103
Sheffield Midland, 63
Sheffield Victoria, 170
Shildon, 187
Silloth, 97
Stirling, 104
Thornton Junction, 76, 82, 108, 159, 160, 192
Torphin, 130
York, 17, 18, 20-22, 26, 28, 31, 33, 34, 37, 45, 58, 164, 165, 170, 174, 176
York Holgate platform, 24, 41, 44, 54

Locomotives (LNER numbering)
42 (F, D22), 57
43A, (D47), 149
45A (45, D47), 149, 150
48A (D47), 150
95 (60095, A3), 96
158 (class '38'), 16
201 (D49), 155
211 (D49/2), 164
217 (D23), 60, 62
217 (62753, D49/2), 161
223 (G1, D23), 59
234 (D49/1), 153
235 (D49/2), 156

Index

245 (D49/1), 153
246 (D49/1 preserved), 191, 192
247 (D49/1), 156
250 (62714, D49/1), 183
255 (62742, D49/2), 161
264 (D49/1), 168, 170
264 (62704, D49/1), 182
265 (62705, D49/1), 160
266 (D49/1), 153
269 (62743, D49/2), 182
270 (D49/1), 167
270 (62708, D49/1), 185
273 (62745, D49/2), 175
274 (62759, D49/2), 179
279 (62760, D49/2), 180
288 ((D49/2) 165
292 (62748), 176
306 (D49/1), 168
307 (62716, D49/1), 160
309 (62717, D49/1), 178
310 (62718, D49/1), 179
327 (D49), 154
329 (62725 D49/1), 174
336 (D49), 154
345 (D49), 89
352 (62726, D49/2), 175, 176
356 (F, D22), 57
359 (62763, D49/2), 180
362 (D49/2), 172
365 (D49/2), 170
365 (D), 158
365 (62768, D), 158, 159
366 (Fletcher 2-4-0), 16
368 (62770, D49/2)
375 (D49/2), 171
376 (D49/2), 170
426 (class '38'), 15, 16
472 (D23), 60
663 (F, D22), 57
676, (G1, D23), 59
707 (D20), 33
779 (F, D22), 56
1026 (D20), 31
1051 (D20), 32
1147 (D20), 38
1207 (D20), 37
1234 (D20), 32
1237 (D21), 50

1238 (61238, B1), 46
1238 (D21), 49, 50
1239 (D21), 50, 54
1241 (D21), 54
1244 (D21), 53
1260 (D20), 42
1324 (D 2-4-0), 55, 58
1356 (61356, B1), 109
1357 (61357, B1), 109
1506 (Tennant 2-4-0), 16
1524 (class '38'), 15
1540 (F, D22), 58
1619 (D19), 29, 30
1620 (D17), 22
1621 (D17 preserved), 186, 187
1629 (D17), 18, 21
1631 (D17), 18
1638 (D17), 17
1870 (D18), 27, 28
1871 (D17), 19
1874 (D17), 19
1901 (D17), 26
1903 (D17), 25
1908 (D17), 24
1923 (D17), 26
2011 (D20), 31
2020 (D20), 33
2021 (D20), 41
2111 (ex 1873 D17), 20, 26
2347 (62347, D20), 48
2349 (62349, D20), 35
2360 (62360, D20), 48
2362 (62362, D20), 36
2366 (62366, D20), 36
2369 (E2369, D20), 34
2370 (D20), 45
2371 (62371, D20), 46
2376 (62376, D20), 36
2383 (62383, D20) 34
2386 (62386, D20), 35
2387 (62387, D20), 47, 48
2392 (D20), 34
2753 (D49/1), 166
2753 (62728, D49/1), 160
2754 (D49/1), 59
2755 (D49/1), 174
2756 (62731, D49/1), 173
2757 (D49/1), 155

2758 (D49/1, 62733), 91
3033 (J, 33), 63
3038 (J, 38), 63
3042 (2429, D24), 64
4698 (K2), 113
6384 (62677, D11/2), 185
6803 (3, D39), 121, 122
6805 (5, D46), 148
6806 (6, D46), 148
6807 (D42, 142
6810 (10, D42), 141-143
6813 (D43), 144, 145
6814 (D43), 145
6819 (D41), 133
6825 (25, D40), 123
6825 (62260, D40), 125
6829 (62267, D40), 131
6833 (D40), 124
6845 (D40), 124, 129
6845 (nameplate, D40), 126
6846 (D40), 128
6847 (62275, D40), 130
6848 (62276, D40), 130, 131
6849 (49 preserved), 189, 190
6850 (50, D40), 123, 124, 128, 129
6851 (51, D45) 147
6852 (62279, D40), 126
6854 (54, D40), 123
6867 (67, D44), 146
6871 (D48), 151
6874 (D42), 142, 143
6875 (75, D38), 120
6877 (77s, D38), 120
6878 (62225, D41), 135, 136, 140
6879 (79, D41), 133
6883 (D41), 134
6883 (62230, D41), 139
6893 (93, D41), 132
6897 (62241, D41), 136
6899 (2243, D41), 135
6902 (62246, D41), 135
6904 (62248, D41), 136
6905 (D41), 134
6905 (2249, D41), 138
6907 (107, D41), 138
6911 (62255, D41), 140
6912 (62256, D41), 136, 139
6913 (62262, D40), 125

6915 (62264, D40), 131
7633 (67633, V1), 114
9149 ((149, D34), 107
9216 (216, D31), 98
9218 (218, D31), 97
9221 (D34), 110
9231 (231, D35), 116
9241 (2481, D34), 108
9242 (D34), 113
9243 (2406, D29), 76
9244 (244, D29), 76
9244 (2407, D29), 76
9245 (D29), 79
9256 (D34 preserved), 188
9278 (62484, D34), 109, 115
9298 (62479, D34), 109
9307 (62472, D34), 114
9319 (319, D26), 69, 70
9320 (320, D26), 70
9333 (62457, D33), 106
9361 (361, D29), 77
9362 (362, D29), 75
9363 (D30), 89
9400 (400, D30), 81.,84
9406 (D34), 108
9406 (62474, D34), 113
9411 (411, D29), 81
9412 (62421, D29), 182
9416 (D30), 82
9416 (62425, D30), 92
9417 (D30), 89
9420 (62429, D30), 84
9422 (2431, D30), 83
9424 (D30), 92
9425 (62434, D30), 93
9426 (426, D30), 82
9427 (62436, D30), 91
9428 (62437, D30), 83
9490 (62489), D34, 114
9494 (62496, D34), 115
9500 (62441, D30), 83

9502 (62490, D34), 109
9503 (D34), 107
9592 (592, D25), 66
9594 (594, D25), 67
9595 (D25), 67
9595 (595, D25), 68
9602 (602, D25), 66
9603 (603, D25), 67
9633 (633, D31), 94, 95, 97
9634 (D31), 98
9635 (D31), 96
9635 (62281, D31), 96, 100
9695 (D36), 118
9732 (732, D31), 95
9734 (D31), 95
9767 (767, D31), 98
9769 (2073, D31), 96
9864 (D33), 104
9866 (D33), 104
9866 (62457, D33), 105
9886 (D32), 102
9890 (62451, D32), 103
9892 (892, D32), 101
9893 (D32), 101
9898 (898, D29), 75
9899 (899) D29, 78
9900 (900, D29), 77, 78
9992 (476, D27), 73
9995 (491, D27). 72
10324 (479, D27), 74

Photographs (Colour)
Locations
Alne, 195
Boat of Garten, 199
Carron, 199
Craigellachie, 198, 200, 202
Crianlarich, 198
Dalmeny, 206
Dawsholm, 202
Elgin, 201

Haymarket West, 197
Humshaugh, 195
Glasgow Eastfield, 196
Kirkham Abbey, 206
Leeds, 194
Northallerton, 194
North Queensferry, 196
Reedsmouth, 205
Whitrope summit, 197
York, 193, 204, 205

Locomotives (number as photographed)
49 (D40 preserved), 202, 203
235 (D49/2), 204
246 (D49/1 preserved), 208
1207 (D20), 193
2759 (D49/1) 204
9035 (D34), 196
61783 (K2), 202
62264 (D40), 201
62271 (D40), 198, 199
62277 D40), 199, 200, 202
62360 (D20), 194
62370 (D20), 193
62387 (D20), 194, 195
62423 (D29), 195
62441 (D30), 196
62471 (D34), 198
62478 (D34), 197
62488 (D34), 197
62496 (D34), 198
62701 (D49/1), 205
62712 (D49/1), 207
62725 (D49/1) 206
62730 (model D49/1), 208
62731 (D49/1), 207
62751 (model D49/2), 208
62756 (D49/2), 206
62771, (D49/2) 205